*… [A]nd as we arrange the sequence of evolution's advance, we discover an unsettling implication.*

*Each step is an evolutionary curve; all steps together outline an accelerating advance for all biological evolution. Over half of the time was used to advance from procaryote to eucaryote. It took half again the time to reach the level of fish. And as the succeeding steps followed, the succession time shortened. It is the curve of an accelerating object building momentum, like a ball dropped from a height. The driving force goes unchecked; momentum sets the pace.*

*Each major development in evolution begins slowly but, fed by its own momentum, begins to accelerate until it races to its developed state. When it reaches a final level—a high stage in evolution—the offspring of the new life form begin to repeat the cycle, evolving some feature that ultimately leads to another succeeding step.*

*Segments of biology equilibrate and stop evolving, but the overall advance of the column does not reach equilibrium. To the contrary, it continues to accelerate stage after stage to such a rate that it suggests that the interval of man's preeminence will be ominously short. We apparently have reached a critical point in biological evolution. Either the trend of evolution is no longer valid or a radical change in the evolutionary process is imminent. In any event, we are in the middle of something momentous that is taking place.*

—**William Day, *Genesis on Planet Earth***
(ch. 28, Yale University Press, 1984)

# The Evolutionary Trajectory

THE WORLD FUTURES GENERAL EVOLUTION STUDIES
A series edited by Ervin Laszlo
The General Evolution Research Group
The Club of Budapest

*VOLUME 1*
**NATURE AND HISTORY: THE EVOLUTIONARY APPROACH FOR SOCIAL SCIENTISTS**
Ignazio Masulli

*VOLUME 2—KEYNOTE VOLUME*
**THE NEW EVOLUTIONARY PARADIGM**
Edited by Ervin Laszlo

*VOLUME 3*
**THE AGE OF BIFURCATION: UNDERSTANDING THE CHANGING WORLD**
Ervin Laszlo

*VOLUME 4*
**COOPERATION: BEYOND THE AGE OF COMPETITION**
Edited by Allan Combs

*VOLUME 5*
**THE EVOLUTION OF COGNITIVE MAPS: NEW PARADIGMS FOR THE TWENTY-FIRST CENTURY**
Edited by Ervin Laszlo and Ignazio Masulli with Robert Artigiani and Vilmos Csányi

*VOLUME 6*
**THE EVOLVING MIND**
Ben Goertzel

*VOLUME 7*
**CHAOS AND THE EVOLVING ECOLOGICAL UNIVERSE**
Sally J. Goerner

*VOLUME 8*
**CONSTRAINTS AND POSSIBILITIES: THE EVOLUTION OF KNOWLEDGE AND THE KNOWLEDGE OF EVOLUTION**
Mauro Ceruti

*VOLUME 9*
**EVOLUTIONARY CHANGE: TOWARD A SYSTEMIC THEORY OF DEVELOPMENT AND MALDEVELOPMENT**
Aron Katsenelinboigen

*VOLUME 10*
**INSTINCT AND REVELATION: REFLECTIONS ON THE ORIGINS OF NUMINOUS PERCEPTION**
Alondra Yvette Oubré

*VOLUME 11*
**THE EVOLUTIONARY TRAJECTORY: THE GROWTH OF INFORMATION IN THE HISTORY AND FUTURE OF EARTH**
Richard L. Coren

This book is part of a series. The publisher will accept continuation orders which may be cancelled at any time and which provide for automatic billing and shipping of each title in the series upon publication. Please write for details.

# The Evolutionary Trajectory

## The Growth of Information in the History and Future of Earth

**Richard L. Coren**

*Drexel University*
*Philadelphia, Pennsylvania*

**Gordon and Breach Publishers**

Australia • Canada • China • France • Germany • India
Japan • Luxembourg • Malaysia • The Netherlands • Russia
Singapore • Switzerland • Thailand

 Printed in India.

Amsteldijk 166
1st Floor
1079 LH Amsterdam
The Netherlands

---

British Library Cataloguing in Publication Data

Coren, Richard L.
The evolutionary trajectory : the growth of information in the history and future of earth. – (World futures general evolution studies ; v. 11)
1. Evolution (Biology)
I. Title
576.8

ISBN 90-5699-601-0

To *Adam, Justin* and *Emily* in the hope that they will grow up to appreciate, with the scholar and the poet, that

*Happy is the one who finds Wisdom,*
*the one who gains understanding.*
*For its fruits are better than silver,*
*its yield than fine gold.*
*It is more precious than rubies;*
*No treasure can match it.*

—Proverbs 3:13ff

# Contents

# Introduction to the Series

*The World Futures General Evolution Studies* series is associated with the journal *World Futures: The Journal of General Evolution.* It provides a venue for monographs and multiauthored book-length works that fall within the scope of the journal. The common focus is the emerging field of general evolutionary theory. Such works, either empirical or practical, deal with the evolutionary perspective innate in the change from the contemporary world to its foreseeable future.

The examination of contemporary world issues benefits from the systematic exploration of the evolutionary perspective. This happens especially when empirical and practical approaches are combined in the effort.

*The World Futures General Evolution Studies* series and journal are the only internationally published forums dedicated to the general evolution paradigms. The series is also the first to publish book-length treatments in this area.

The editor hopes that the readership will expand across disciplines where scholars from new fields will contribute books that propose general evolution theory in novel contexts.

# Preface

This book presents an entirely novel understanding of changing systems in general and of "evolution" on the earth, in particular. This is done through the development of three concepts, each new and, therefore, controversial.

First is a cybernetic model called "evolutionary escalation." Like standard logistic theory from which it springs, the formal simplicity of escalation is belied by the broad applicability it appears to have.

Second is the trajectory of evolution on earth which, though suggested by the escalation model, is ultimately independent of it—having an empirical reality of its own. This demonstrates the continuity of a single evolutionary process extending from the cosmic through the biological to the technological. Many fine points may be found on which to raise objections to this trajectory; its acceptance requires a temporary suspension of some traditional prejudices so that the relation can be acknowledged in its own right. If this is done, a new world view, a *weltanschauung* if you will, emerges from the coherence it lends to evolutionary change.

Third is the inference from points on the trajectory that information, in its various forms, is the underlying evolutionary variable. Though long acknowledged to be of major significance, this is the first time that its role has been specifically demonstrated.

In addition, we extrapolate the trajectory into the imminent future. The timing of such an extension seems undeniable and we discuss its possible forms.

The book is directed to two professional audiences: one involved in the study of general systems and the other in evolution, as well as to scientific lay readers. I feel that the analysis and results are exciting enough and of general enough interest that they should be presented to the cognizant general public. As a result, the few mathematical relations are explained and clarified, and the mathematical passages are presented in such a way that they may be read without much difficulty. Introductory discussions cover the mathematical, physical, biological and evolutionary backgrounds necessary to appreciate the new results. I have tried to avoid the technical jargon found in professional discussions. Such terminology often carries refined connotations that will thereby be lost, but this is a small price to pay for the potentially expanded readership.

And this is done not only for the lay reader but also for the professional. The range of topics discussed is broad enough that a specialist in one field may not be familiar with the sophisticated concepts and idioms of another field. The Evolutionary Trajectory developed here calls for an appreciation of the developments of cosmology, biology, the very basis of life, sapience, sociology and technology. I certainly cannot claim expertise in these diverse areas; I have spent considerable time and had a great deal of enjoyment becoming passably knowledgeable in them, enough to present and justify their relevance to the trajectory.

Analyses of most evolutionary processes, such as progressive social change—and I make a claim for the attention of sociologists/historians as well—or the theory of biological evolution, are hampered by the lack of comprehensive, quantitative, phenomenological descriptions. In other sciences, understanding of mechanisms and relations has grown out of such phenomenology. It is my belief that The Evolutionary Trajectory presents such a description. We will see examples of this from several areas of human and natural development. And while the trajectory is inherently quantitative, because of its simplicity it can be understood in qualitative terms. Once this is done, the reader can be no more surprised than I by the wide range of general understanding and specific conclusions that can be drawn.

# Acknowledgments

Thanks are due to Chris Rorres for directing me to the mathematical description of escalation, and to Jerry Chandler and the Washington Evolutionary Systems Society for their encouragement at a time when I badly needed it.

Only Elaine and I know how much we both put up with during the years this work struggled into existence.

# INTRODUCTION

This book is about the rate of evolutionary change on earth, how that change is measured and interpreted, cyclic features of its occurrence, and its continuity to the present moment (and beyond). Its main purpose is to show that critical evolutionary events conform to a quantitative, empirical relation, the Evolutionary Trajectory, that brings coherence to many of their aspects that have heretofore been regarded as, at most, analogous. Perusal of the pertinent literature reveals that the study of evolution has taken two distinct directions, with very little overlap and communication between those developing each. These are the analyses through cybernetics and the studies through paleontology/biology.

Cybernetics is relatively recent; its original scope is indicated by the title of the 1961 germinal book[1] by Norbert Wiener: *Cybernetics or Control and Communication in the Animal and the Machine.* This deals with the relations between structure, interactions, and behavior in the living organism and in the machine, their common features, and their description. Although the philosophical consequences are dealt with in detail, the main impact of Wiener's work has been through development of mathematical models of behavior and interaction, utilizing engineering methods of general systems control. Since its formulation, the understanding of what constitutes cybernetics has expanded and only shortly thereafter Turchin[2] pointed out that it had become "the science of relationships, control, and organization in all types of objects. Cybernetics concepts describe physicochemical, biological, social, and technological phenomena with equal success."

This breadth is, perhaps, one reason for the slow acceptability of cybernetics. While we readily accept that there are mathematically formulated laws governing the behavior of physical systems and aspects of biological systems, we are reticent to acknowledge that the same principles may describe our own cumulative behavior. Our "free will" seems threatened by the possibility that some of our group behavior and institutions conform to deterministic, or at least to probabilistic predictive relations. This is particularly true in those areas where such considerations have not been previously applied.

The *development* of the system state, as opposed to its *description* by control models, is not ordinarily considered part of the cybernetic domain, though changes of social structures and society have received a fair amount of attention. However, this, and the fact that living systems are but one aspect of cybernetic consideration (and a difficult, complex one at that), makes it understandable that attention to evolution has been a neglected area of its development. One recently A. Jdanko[3] pointed out that "Two main shortcomings seem to be characteristic of Cybernetics...: First, a neglect of physical entropy and, therefore, of ... information; ... and second, a lack of an evolutionary perspective ...." He and others have published proposals toward the development of an Evolutionary Cybernetic Systems Theory.

Those engaged in paleontology and the biological study of evolution deal with its issues and questions from the points of view of microevolution and macroevolution, terms that have had and now have more than one meaning each. We can take microevolution to comprise the detailed descriptions, mechanism and changes that occur to the members of a class of creatures or of a clade (a group of species having a common ancestry). An example is allometry, the study and measurement of changes such as of body shape and length. One result of such studies is Cope's rule, that body size tends to increase in the course of evolutionary periods.

We can understand macroevolution as dealing with more general understanding of the mechanisms of change. In the opening chapter of the book *Patterns of Evolution*[4] S.J. Gould poses three basic macroevolutionary questions that paleontologists have asked about the history of life. These are:

"(1) Does the history of life have definite directions; does time have an arrow specified by some vectorial property of the organic world ...?" By this he intends to make the following distinction: Organic evolution may be a steady-state phenomenon, meaning that changes occur at random, producing a wide variety of creatures and properties, with some few variants that prosper and grow. Alternatively, it may be directional, meaning that the changes that occur are part of a subtle tendency toward enhancement of some characteristic such as improved efficiency or greater complexity.

"(2) What is the motor of organic change? More specifically, how are life and earth related?" This question is interpreted to ask whether the environment controls evolution, with the changes of various species and families being merely responses to external forces, or are evolutionary changes internally driven, with living things influencing both their own developments and those of the environment, to which they respond?

"(3) What is the tempo of organic change? Does it proceed gradually in a continuous and steady fashion, or is it episodic?" A gradual evolutionary process is one in which the sum of incremental changes eventually produces a total change that is great enough to be considered significantly different from its original form. By contrast, an episodic, or punctuational, scenario implies that long periods of relatively unchanged stasis are punctuated by shorter intervals of rapid change?

Gould points out that his three questions are not new. However, in the last quarter century there has been an accumulation of tools and knowledge applied to them, as well as a new sophistication of relevant models and concepts. Scholars have only recently been able to offer more definite answers than were possible over 100 years ago.

Here our opening chapter is a discussion of how time intervals are measured and expressed for the long periods involved in geologic history. The next three chapters review the pertinent changes that took place in the cosmic formation of earth, in biological evolution on earth, and as a result of the works of mankind. We then step back and discuss some of the ways in which growth and change are described in complex systems. For this, in chapter 6, we extend and develop logistics, a well-known cybernetic growth model, to include the phenomenon of "escalation," i.e., sequential, punctuated evolution. As a result we can apply *quantitative* analysis to evolutionary changes; chapter 6 derives The Evolutionary Trajectory on Earth, a phenomenological relation describing the progress of evolution. The trajectory calls for a sequence of dates of the major "transition" events of evolution, and these are drawn from the earlier discussions, using widely accepted classifications and criteria, and based on the judgements of scholars in their own areas of expertise.

The existence and nature of The Evolutionary Trajectory expands the concept and extent of evolution beyond what is usually accepted or expected. Chapters 7 and 8 discuss the basis of coherence of this new representation and we respond to a modified form of Gould's question 2. Instead of "What is the *motor* [i.e., mechanism] of organic change?" we ask what is the *measure* of evolutionary change? The answer is entropy or, preferably, its complement: information. A novel interpretation, a *weltauschauuing*, if you will, emerges from the unity of the series logistic cycles, and some of its philosophical implications are considered in chapter 9. There we find that information is also the answer to a modified form of question 1: Does evolutionary time have "an arrow?" The meaning of the answer to this important inquiry demands careful examination. The series of emergent logistic cycles also provides a response to Gould's third question of punctuated or graded evolution, but not on the micro or macro-evolutionary level. Instead, on a "super-macroevolutionary" level The Evolutionary Trajectory implies cyclic, non-uniform development.

That information enters all aspects of evolution means that the criteria for defining organic evolutionary stages are not exclusively biological. This is not an entirely novel idea for biologists but, nevertheless, it requires initial suspension of inbred and often strongly held beliefs. In fact, The Trajectory demonstrates that evolution has already left the organic mode in favor of a technological one. Chapter 10 discusses the consequences of this change and of the reduced timing implied by The Trajectory. One consequence of the apparent directionality of The Evolutionary Trajectory is that its extrapolation into the future indicates, in the words of the fronticepiece, that "we are in the middle of something momentous that is taking place."

# 1: QUANTITATIVE MEASURE

## A New Understanding

In common usage, earthly evolution is generally understood to be the series of changes from simpler to more complex, or from "lower" to "higher" biological forms, that have produced the present make-up of life on our Earth. Before we even begin to examine the details of those changes we should recognize that merely to use the word "evolution" in this context acknowledges that change occurs over very great time periods. Awareness of evolution is so common today that it may surprise a modern reader to learn that recognition of the occurrence of long term change is relatively new. Until quite recently, historically speaking, it had been universally assumed that the earth and its host of living things were unaltered since an initial creation, except for the events of recorded or spoken history. And, in Western belief, the total elapsed time since the seven days of The Creation in biblical Genesis was, arguably, quite short. For example, counting back the various genealogies given there, Archbishop James Ussher (1581–1656) established[1] the total time since creation such that the common year 2000 will be 6004 *anno mundi.* With somewhat different reckoning it will be 5760 of the Hebrew calendar.

The modern roots of departure from this belief in the immutability of events and things can probably be traced to the sixteenth and seventeenth centuries religious Reformation and counter-Reformation, and the tremendously destructive wars they wrought in central Europe. The result of those events was to end the centuries old control of the Roman Church over all intellectual, social, religious, and political matters. Among other changes, they released the economic and intellectual fire, known as the Enlightenment, that had been smoldering at least since the Crusades of the eleventh, twelfth, and thirteenth centuries. It had grown even hotter following the invention of rapid printing in the fifteenth century and was now able to spread through the Western world. Those crusades, the subsequent commerce, and the general availability of printed works exposed Europeans to modes of thought and learning which, during the European dark ages, had been kept alive only in some monasteries and universities of Europe and in Arabian courts. In particular, the study of natural phenomena slowly penetrated centers of scholarship. Galileo's trial for heresy in 1632 is but one example of the struggle between the old orthodoxy and the new awareness. In spite of its ability to punish him, the power of the Church was diminished. By the time of Darwin's *Origin of Species* in 1859 the defense of orthodoxy was largely intellectual rather than blasphemous, although surely based on emotional and religious foundations. While this battle is still being fought in some places, its very existence admits of a new, secular understanding.[2]

## Geological Time

These changes allowed speculation on the interpretation of what were otherwise incoherent facts, such as the existence of fossils. These were noted 2600 years ago by Xanophanes, and regarded as clues to earthly change. In the fifteenth century, Leonardo da Vinci pointed out that oceanic fossils on high mountains could not be ascribed to the biblical deluge. He noted that natural processes that are even presently changing the earth, such as weather erosion, inundations, and river deposits, are sufficient explanation. Like many of his other works, Leonardo's idea never gained popularity, though it appeared from time-to-time in the following centuries. Then, in 1785, James Hutton, in his book *The Theory of the Earth,* used examples from his native Scotland and from continental Europe to reestablish this idea, as a "Principle of Uniform Causes." He also proposed geologic "Superposition," that deeper undisturbed rock layers are older than those that lie above.

Just before 1800, William Smith carried these proposals forward to the point that he could assign ages to the various rock layers, or strata. He noted that each sedimentary rock formation contains distinctive types of fossils that change between the strata. When the same fossils are found in rocks at the distant sites, they must represent the same geologic period, and he was able to put the various layers, from different sites, in time sequence. Of course, the same principle applies to fossils of aquatic and land flora and fauna, found in intervening layers and to the evidence of human activity found in the highest, most recent layers. By studying present physical, geologic processes, the times of formation of the layers were estimated (for example, the process of depositing fine slit from a river bed is slower than that of building layers of coarser sand on an ocean floor) so that a time scale could be assigned to the different strata. In this way, the fossils sequence the rocks and the rocks date the fossils. However, because of other intervening processes, such as volcanism, metamorphism, mountain building, and erosion, and because of uncertainties of the true rates of geologic events, this dating is relative and, until recently, absolute dates were highly uncertain.[3]

Despite the vast gathering of evidence of geologic change, through biological taxonomy and related geological classifications and orderings of rock strata and fossils, supposedly knowledgeable people held to the old beliefs until they were clearly untenable. "Even in the middle years of the nineteenth century it was seriously contended that fossils, suggesting as they did another story, had been hidden in the ground by God (or perhaps the Devil) to test the faith of man."[Dampier[1]] It is even more surprising to find this still expressed in 1989.[4]

There are now available irrefutable and more precise measures of the ages of rocks. Before discussing them, however, we should be aware that geologists and paleontologists divide geologic time into intervals, depending on the characteristic rocks and the distinctive life forms associated with them. The nomenclature for these divisions developed initially in a somewhat haphazard manner, with different names being assigned according to local site names or rock appearances, and frequently having different bases for making divisions.[3] Although there is still some variation in naming, e.g., between lesser time intervals used in North America and in Europe, general agreement has been reached on most of the major units of time. In order of decreasing duration these are divided into

Eons
Eras
Systems
Periods
Epochs
Ages

In the following we will limit our discussion to those divisions needed later. From oldest to youngest, the specific divisions are:

¤ The Precambrian Eon

- The Azoic, the earliest era, is characterized by the meaning of its name: without life. During this interval, the crust of the earth repeatedly melted and hardened until a permanent cover developed, and the oceans and an atmosphere were formed. This era begins with the birth of the earth and lasts until the first signs of microscopic fossil life are found.

- The Archeozoic Era. Its name means the oldest or first time of life and its fossils are of bacteria and algae, the most primitive forms of life. This was still a time of great volcanism, rock flow, and mountain building. The Azoic and Archeozoic are sometimes grouped together as a single entity.
- The Proterozoic Era is marked by the appearance of simple invertebrate animals (i.e. without bones). Widespread glaciation took place along with great floodings and land mass motion.

¤ The Phanerozoic (visible life) Eon
This opens with the Paleozoic era, whose first subdivision is the Cambrian Period. There is an abrupt appearance of abundant fossils. It is this important demarcation that gives rise to the division of the geologic record into Precambrian and Phanerozoic Eons. Although there are commonly assigned many more divisions to the Phanerozoic Eon than to the Precambrian, that is because the subsequent development of life allows finer distinctions between its various stages. In fact, the Precambrian Eon lasted about eight times longer. The eras of the Phanerozoic are

- The Paleozoic Era. As already mentioned, the rocks of this era contain fossils of simple shelled animals, fish, reptiles, and the first forests. This era is constituted of seven periods.
- The Mesozoic Era includes the rise and fall of Dinosaurs and saw the development of birds and of modern plants. It consists of three periods.
- The Cenozoic Era is the most recent. It is divided into two or three Systems, depending on the intent of the categorization:
    1. The Paleogene System witnessed the development of mammals and other modern types of animals. Sea flooding and mountain building continued, particularly the rise of the Alps at the end of this System, making it a natural dividing point for European geologists. This interval is also biologically characterized by the presence of Nummulites, large, shelled, one-celled creatures. The Paleogene is further divided into three epochs.
    2. The Neogene System. At this point the epochs of the system are sufficiently short and involve sufficiently detailed fossil evi-

dence that we should refer to them individually. The Neogene is divided into two epochs:

i – The Miocene
ii – The Pliocene

These involved more specializations and divergences of cladistic lines. They were also times of great geographic and climatic change.

3. The Quarternary is the most recent time division, so that there is abundant fossil evidence for some of its aspects. It encompasses human appearance and development so it has been intensely studied. Among scholars, the Quarternary is sometimes considered to be a System, following the Paleogene and Neogene; sometimes it is listed as a Period; and sometimes as an epoch. The exact category need not involve us. Its epochs and ages are:

   i – The Pleistocene Epoch is divided into three ages:
   (a) Lower Pleistocene
   (b) Middle Pleistocene
   (c) Upper Pleistocene

   based on the series of great glaciations that filled its time.[5] It is distinguished by the appearance of manlike creatures whose stages of development parallel its three ages,

   ii – The Holocene or Recent Epoch is the modern age

It was noted above that the geologic divisions become shorter and more detailed as the present is approached. Still their durations are measured in millions of years and, although the relative ages of the rocks and life types could be placed according to the Principle of Superposition, as listed above, absolute ages were grossly uncertain. More accurate dating depended on the development, particularly within the last 50 years, of the techniques of radioisotope determination, and analysis of the times for nuclear decay.

Radioactive materials are elements whose atomic nuclei disintegrate spontaneously, causing them to radiate electromagnetic energy and to emit subatomic particles and leaving nuclei corresponding to other elements. To understand the method of radio-dating and to enable us to compare and deal with the vast geologic time periods involved, we must become familiar with scientific notation for stating numbers. This lends itself particularly well to measuring quantities that vary over very great ranges of size, time, activity, etc. We will then be able to examine the method of establishing and expressing geologic time.

## Orders of Magnitude

In mature sciences, such as physics, chemistry, modern biology, etc., measurements are made and the results presented in the form of numerical dimensions, e.g., so many kilometers or microseconds. A problem can arise when comparing different measures because of he tremendous ranges encompassed by some of these dimensions. Physical sizes range from less than that of an atomic nucleus, about a millionth of a billionth of a meter, through man, about 1 to 2 meters, through the distance to the nearest star, more than 10 billion billion meters, to even greater galactic and cosmic sizes. Similarly, measures of time range from the duration of high energy, nuclear phenomena, which transpire in intervals of a billionth of a billionth of a billionth of a second, and even less, through our own life spans of under a hundred years, or several billion seconds, to the age of the universe, presently estimated at greater than ten billion years, or a billion billion seconds.

It is immediately seen that the use of billions and millionths is awkward. It is difficult to compare physical lengths or time durations, that is, to state just how much bigger or smaller (or older or younger) one thing is than another. Furthermore there is not even agreement on the use of some quantitative names. In the United States a billion means a thousand million (a one with nine zeros: 1,000,000,000) while in most other countries it is a million million (a one with 12 zeros: 1,000,000,000,000); this is denoted by a trillion in the U.S. And there are other dimensional names that are likely to be used in one culture and not in another, for example, the common measure: one milliard (one thousand million = one U.S. billion), is rarely used in U.S. Therefore, a simple, universal system of designation is needed, and it is must be one that is capable of covering the vast dimensional ranges. For this the concept of orders of magnitude is most useful.

We will take the term "order of magnitude" (which we abbreviate by OM) to be simply a statement of the number of zeros that accompany unity in characterizing the dimension. Thus, 1000 seconds has three zeros so its OM is 3. Because 1000 is equal to $10 \times 10 \times 10$, i.e., 10 multiplied by itself three times, it is written as $10^3$, stated as being ten to the third power; the OM appears as the exponent of 10. From this we see that one million, $1{,}000{,}000 = 10^6$, has OM = 6, and there is no ambiguity if a billion is stated to be $10^9$ (OM = 9). With this system we have that ten, which has a single zero after the 1 is $10 = 10^1$ (OM = 1), and unity, which has no zeros, is $10^0 = 1$ (OM = 0).

Note also that the OM is the number of places, or zeros, through which the decimal point must be moved to the right from unity: 1.000..... This is a useful observation when considering measures smaller than unity. One hundredth, written in decimal form as 0.01, involves moving the point to the left through two places from 1.0; its order of magnitude is –2, where the negative denotes the opposite direction to shift the decimal point. In powers of ten this is $10^{-2}$. One hundredth can also be expressed in fraction form as $1/100 = 1/10^2$, demonstrating that a negative OM corresponds to a positive power placed in the denominator of a fraction with unity numerator. Thus, one microsecond (one millionth of a second) is $0.000\ 001 = 1/1{,}000{,}000 = 1/10^6 = 10^{-6}$ seconds (OM = –6).

It is, perhaps, appropriate to indicate another problem that is alleviated by the use of OM, that of decimal notation itself. In common European usage, one writes one millionth as $10^{-6}$ = 0,000 001 where the decimal location is denoted by a comma (,) and the zeros are commonly spaced in units of three. In North America this is expressed $10^{-6}$ = 0.000001 where the decimal location is denoted by a period (.) and spacing for clarity is rare. In either convention, there is no ambiguity in writing $10^{-6}$ or OM = –6.

Powers of ten are extremely convenient when comparing sizes because exponents are added when multiplying numbers. Thus, one thousandth of a second, a millisecond, is $0.001 = 10^{-3}$ sec, so that 100 milliseconds is $100 \times 10^{-3} = 10^2 \times 10^{-3} = 10^{2-3} = 10^{-1} = 0.1$, a tenth of a second. Similarly, 1000 times 100 milliseconds is $10^3 \times 10^{-1}$ sec. $= 10^2 = 100$ sec. As another example, consider that we wish to mark off one per cent of the distance between two places separated by 1 kilometer. Since one per cent is one hundredth $= 10^{-2}$, and 1 kilometer $= 10^3$ meters, we have one per cent of 1 kilometer $= 10^{-2} \times 10^3 = 10^{-2+3} = 10^{-1} = 10$ meters. This answer could also have been obtained by observing that powers of ten are subtracted when dividing: 1 per cent of 1 kilometer $= 1/100 \times 10^3 = 10^3/10^2 = 10^{3-2} = 10^1 =$ 10 meters.

The practical use of powers of ten, in nature, is illustrated by our physical senses, which have adapted us for survival through the use of OMs.[6] We prevent being overwhelmed by high light intensities by reducing the size of the iris of the eye and the sensitivity of our visual receptors and neural transmitters. These measures extend the intensity range of useful vision, yielding the ability to see in very low and in very high light levels. In a similar way, there are muscles in the ear that adjust the tension of the ear drum, the motion of the middle ear ossicles, and the stiffness in the inner ear. As a result, very low sound levels can be detected and interpreted in one circumstance while, in another, very intense levels can be received without damaging the ability to hear. Because of these and other adaptations, our visual and auditory responses, i.e., our *subjective senses* of brightness or loudness, are not in proportion to the physical intensities of the light or sound. They are proportional to the orders of magnitude of those intensities.

Sound waves are pressure fluctuations of the air that cause our ear drums to vibrate. At the frequency of middle C on the piano, the lowest audible sound has an intensity of about $I_o = 10^{-11}$ Watts per square meter. Table 1-1 lists six typical sounds[6,7] along with their absolute intensities, their relative intensities, and the OM's of the relative intensities (rounding the numerical values). If our subjective sense were proportional to intensity then each successive sound would seem to be about 100 times louder than that before it. If we were able to hear the lowest we would find that the loudest, which is more than $10^{10}$ times as intense, exceeds our ability to respond, and our auditory mechanism might even be damaged. However, this is not the case; our sense of loudness is proportional to the OM of the sound. Since the differences of OM of each of these sounds are 2, the third will only seem, subjectively, as much louder than the second as the second does relative to the first, and the same for the fourth to third, etc. This adaptation, by means of OMs, enables us to tolerate and comprehend a tremendous range of stimuli.

OM is also convenient for the presentation of numbers in a compact way, without losing their true magnitudes. For example, we can state that

**TABLE 1-1**
APPROXIMATE SOUND LEVELS (256 HZ.)

| SOUND | INTENSITY (I) $W/m^2$ | $I/I_o$ | $OM(I/I_o)$ |
|---|---|---|---|
| Threshold of Hearing ($I_o$) | $10^{-11}$ | 1 | 0 |
| Whisper | $8 \times 10^{-10}$ | $0.8\ (10)^2$ | 2 |
| Soft Music | $2 \times 10^{-7}$ | $2\ (10)^4$ | 4 |
| Conversation | $10^{-5}$ | $10^6$ | 6 |
| Orchestra | $9 \times 10^{-4}$ | $0.9\ (10)^8$ | 8 |
| Road Construction | 0.25 | $2.5\ (10)^{10}$ | 10 |

the speed of light in vacuum is 299,792,458. meters per second, or we can put this in the form 2.997 924 58$(10)^8$ m/s. Both of these forms have nine significant figures, i.e., decimal place filled with a meaningful value. Such a long sequence of known digits tells us that the speed of light is known with very great precision. If we only wish to retain four significant figures we have 2.998$(10)^8$ m/s. Note that the power of 10 is the same with this change; the order of magnitude of its size is independent of the precision with which a quantity is measured or stated. In this case we frequently write $3(10)^8$ m/s, where we have rounded up the fourth digit.

OM is often used alone, i.e., without any multipliers. In such cases it is understood to be only a gross measure of size, ignoring factors of 2 to 3 in the actual dimension. Thus, the statement that the age of the earth is of the order (of magnitude) of $10^{10}$ years is meant to imply is that the age is significantly greater than $10^9$ years and significantly less than $10^{11}$ years. It can be taken to mean that the age lies somewhere between, say, one third or one half of this value to two or three times this value. The distance from New York to San Francisco is about 2000 miles whereas the radius of the earth is 4000 miles; they are of the same OM. Similar "rounding of" was carried out for some values of $I/I_o$ on Table 1-1.

On the other hand, consider that the distance from the earth to the moon is $4(10)^5$ kilometers and that from the earth to the sun is about $1.5(10)^8$ km. The ratio of their distances is significant as it must be nearly the same as the ratio of their diameters if we are to have solar eclipses. This ratio is 375, which can be written as $3.75(10)^2$ or as $0.375(10)^3$; it seems to have an OM of 2 in the first case and 3 in the second. We see from this that gross estimation can be inadequate in some instances and a more refined measure of OM in needed.

## Logarithms

The discussion so far has dealt only with integer orders of magnitude. To address the issue of finding a more precise definition let us turn to non-integer powers of ten; we begin with some specific examples. For the first we ask what is meant by the expression $10^{1/2}$. This can be understood through the multiplication rule, described above. If $10^{1/2}$ equals some number "n" then, if we multiply n by itself we have $n \times n = 10^{1/2} \times 10^{1/2} = 10^1$, because exponents add in multiplication. Now some numerical trials will soon reveal that if n = 3.162 then multiplying it by itself yields (very close to) 10. Therefore $10^{1/2} = 10^{0.5} = 3.126$. In a similar way we can find that $10^{1/3} = 10^{0.3333} = 2.154$ since $2.154 \times 2.154 \times 2.154 = 9.9939$, which is very close to 10. (More trials with a calculator can make this value as accurate, as desired.) The quantity $m = 10^{2/3}$ can be found from the fact that $m \times m \times m = m^3 = (10^{2/3})^3 = 10^2 = 100$, leading to the value m = 4.642.

$$10^{2/3} = 10^{0.6667} = 4.642$$

We can infer from these examples that all number between 1 and 10 have OM's, or powers of ten, between 0 ($10^0 = 1$ so the exponent producing 1 is zero) and one ($10^1 = 10$ so the exponent producing ten in unity). The examples given above could be simply worked because the powers of ten were rational fractions (the ratio of two whole numbers). While not all numbers can be expressed in this way any number can, at least, be closely squeezed between two such numbers, so that the appropriate power of ten could be found to any desired degree of accuracy. However this is not necessary as mathematicians have developed faster and more systematic techniques for finding any power of ten that is less than unity. Modern calculators produce these value in an instant. The exact OM of any number, the power of ten producing that number, is called its logarithm (abbreviated "log"): Table 1-2 shows logarithms of the integer from 1 to 10.

TABLE 1-2

| n | log n |
|---|---|
| 10 | 1.000 |
| 9 | 0.954 |
| 8 | 0.903 |
| 7 | 0.854 |
| 6 | 0.778 |
| 5 | 0.699 |
| 4 | 0.602 |
| 3 | 0.477 |
| 2 | 0.301 |
| 1 | 0.000 |

FIGURE 1-1

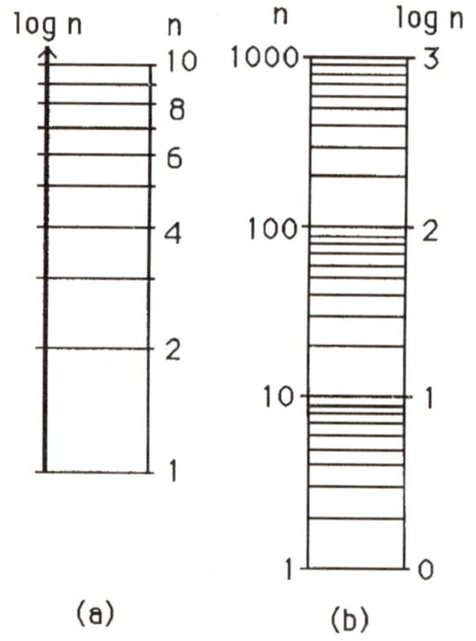


**Table 1-2 and Figure 1-1** Logarithmic scale.

A useful scale can be constructed by marking lengths proportional to the logs of numbers, instead of the number themselves. This is shown, for the integers, by the scale on Figure 1-1a. For example, log4 = 0.602 so the line denoting 4 is placed 60.2% up the scale between 1 and 10.

The integer powers, representing numbers like 100, 1000, etc., and the power for numbers from 1 and 10 can be combined to find the exact power of any number, by adding exponents. For example, 400 = $4 \times 10^2$, so that log 400 = log4 + log 100 = 2.602. As another example, consider that 215.4 must have OM = 2.3333, since we found in the previous paragraph that 215.4 = 2.154 × 100 = $10^{0.333} \times 10^2 = 10^{2.333}$. Similarly, from our other examples: 46.42 = 4.642 × 10 = $10^{1.6667}$ so that 1.6667 = log 46.42, and similarly: 3.5 = log 3162. We see that the logarithms of numbers greater than 10 differ from those between 1 and 10 only by the constant integer preceding the decimal point, without altering their relative placement within the basic ranges. As a result, a scale such as that shown in Figure 1-1b, can represent logarithms of all numbers by merely shifting the scale of Figure 1-1a by the appropriate integer amounts.[A]

Returning to the ratio of distance of the sun and moon, that initiated our discussion of logarithms, we have log 375 = 2.574. We now see that our earlier uncertainty about the OM of this ratio arose from the fact that its logarithm (that is, its precisely determined OM) lies nearly half way between 2 and 3.

Logarithms are extremely useful in describing a whole host of natural phenomena. To mention just a few:

- The acidity or alkalinity of a solution, called the potential of Hydrogen (abbreviated pH) is equal to (–log) of the hydrogen concentration in the solution.
- The Richter scale, used by earth scientists, assigns a measure of earthquake intensity that is proportional to the logarithm of the energy released by the quake. This means, for example, that an earthquake with a scale value of 6 is 100 times (two orders of magnitude) stronger than one with a scale value of 4.
- As mentioned earlier, the perceived loudness of a sound is proportional to the logarithm of its intensity. This logarithmic relation in hearing is not without limits, however. Table 1-1 gave values at a sound frequency (pitch) of 512 Hz; it is found that the minimum discernible sound level depends on the frequency of the sound, for

pure tones, or on the tonal make-up of the sound. This minimum threshold of hearing increases very rapidly from its midrange value, meaning that low intensity hearing becomes poorer, as the frequency is lowered toward 20 Hz or raised to 20,000 Hz. Similarly, at the upper limit of intensities there are also departures from the logarithmic scale. As the sound level approaches $10^{11}$ to $10^{12}$ times the threshold minimum, one's hearing mechanism can no longer adapt, and such extremely loud sounds produce intense pain.

- Another property of hearing is the subjective sense of pitch. Over most of the range of hearing, sounds that differ in frequency by a factor of 10 *seem* to differ in pitch by a factor of two, i.e., subjective pitch is proportional the logarithm of frequency.

In analyzing such phenomena and properties it is often expedient to represent their behaviors graphically. Figure 1-2 is a graph showing three values taken Table 1-1. The relative physical intensity of sound ($I/I_o$) is on the vertical scale and the value of its subjectives loudness is on the horizontal scale. The entire Table 1-1 is presented in modified form, as

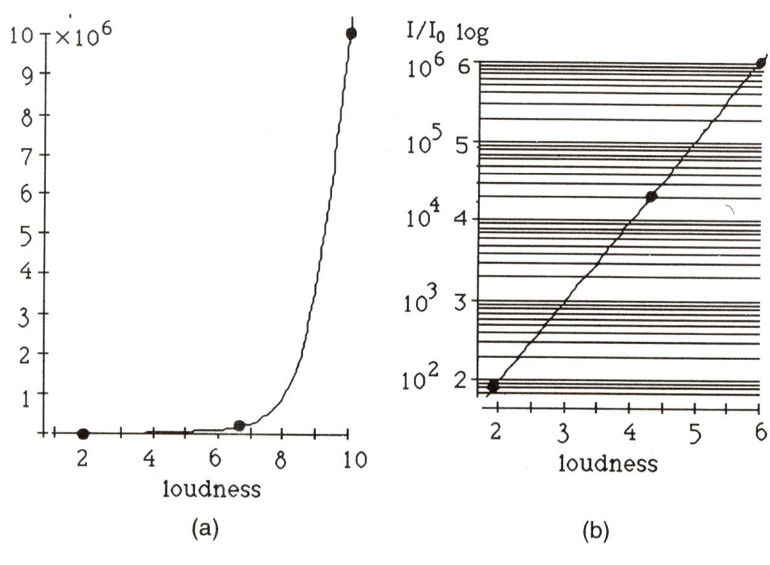


**Figure 1-2** (a) Linear plot and (b) Semi-log plot of sound intensity data.

TABLE 1-3
APPROXIMATE SOUND LEVELS (256 HZ.)

| SOUND | INTENSITY I W/m$^2$ | $I/I_o$ | $\log(I/I_o)$ |
|---|---|---|---|
| Threshold of Hearing ($I_o$) | $10^{-11}$ | 1 | 0 |
| Whisper | $8 \times 10^{-10}$ | 80 | 1.90 |
| Soft Music | $2 \times 10^{-7}$ | $2\ (10)^4$ | 4.30 |
| Conversation | $10^{-5}$ | $(10)^6$ | 6.00 |
| Orchestra | $9 \times 10^{-4}$ | $9\ (10)^7$ | 7.95 |
| Road Construction | 0.25 | $2.5\ (10)^{10}$ | 10.40 |

Table 1-3; the modification is in the last column, giving the subjective loudness as $\log(I/I_o)$. The graphical representation of Figure 1-2a is unsatisfactory because the rapid rise of the curve would require an extremely long vertical scale to show all the points from the table. In addition, with this scale it is difficult to distinguish points with low values, and the curve drawn through the points does not make clear that the abscissa (horizontal axis) values are logarithms of the ordinate (vertical scale) values, rather than having some other dependence.

To overcome these difficulties it is common practice to substitute a scale based on the logarithm of the ordinate value, i.e., the logarithmic scale of values from Figure 1-1, instead of the value itself. This is called a semi-logarithmic plot; it is shown in Figure 1-2b with the same three points. With the new ordinate scale distance are proportional to the logarithms of the numbers being plotted, not to the numbers themselves. We see that those points at low intensities can now be distinguished, as well as those at higher values. Most importantly, the logarithmic plot causes the points to fall on a straight line, clearly demonstrating their functional, logarithmic relationship. This figure could easily be extended to include the other points on Table 1-3. A straight line on a semi-logarithmic graph is a powerful means of establishing that a relation exists between one quantity and the OM (logarithm) of another. In a later chapter we will show that it can also be used to establish an "exponential" relation between two variables since the exponential and logarithm operations are inverse of each other.

## Radio-Dating

We can now return to our earlier discussion of techniques for establishing the absolute dates of geologic and evolutionary events, particularly using nuclear decay of radioactive materials.

In 1895 Wilhelm Konrad Roentgen discovered that X-rays were emitted from gas tubes used for electrical discharge. This initiated a great search for other sources and only a year later Henri Becquerel found that uranium and its compounds also emit penetrating rays. Within the following few years the properties of several spontaneous nuclear emissions were established, although it took several more years until their significance was realized.

Uranium has the highest atomic number of any element found in nature. Its atomic number is 92, meaning that it has 92 positively charged protons in its nucleus, and that nucleus is normally surrounded by a corresponding cloud of 92 negatively charged electrons. More than ninety-nine per cent of uranium atoms have a nuclear mass of 238 and are denoted as $U^{238}$. This means that their nuclei also contain 238–92 = 146 uncharged particles, called neutrons. There are two other isotopes: $U^{235}$ and $U^{234}$. The uranium nucleus is unstable and may spontaneously change, or "decay." In doing so it emits either an alpha particle ($\alpha$) consisting of two protons and two neutrons, or beta particle ($\beta$) which is an electron, or a gamma ray ($\gamma$) which is a high energy electromagnetic X-ray, or it may simultaneously emit more than one of these. $\alpha$ and $\beta$ emission change the uranium nucleus into the nucleus of another element, depending upon its altered atomic number. For example, the emission of an alpha particle depletes $U^{238}$ of two protons, leaving it with 90. The element thorium normally has 90 protons in its nucleus so the altered uranium has become thorium. Since the nucleus also how has 144 neutrons for a total of 234 nucleons, the radio-decay process has produced $Th^{234}$. This daughter nucleus also displays spontaneous radiation activity, i.e., it is also "radioactive." In this way the initial decay has initiated a series of steps, starting with $U^{238}$ and ending with a nucleus of lead, $Pb^{206}$ or $Pb^{207}$ (the most common isotope).

The decay of any particular nucleus is a purely random event and is very unlikely to occur; measurements reveal that a period of $4.5(10)^9$ years must elapse for half of the $U^{238}$ atoms in any collection to decay to $Pb^{206}$. This number is called the half-life of the decay process. From this there is a simple relation between the atomic ratio of $Pb^{206}/U^{238}$ found in a rock sample and the time it must have been decaying, i.e., its age.

Uranium is fairly widely distributed in the earth. It makes up an average of 5 parts per million of rock and is even more concentrated in some minerals, such as pitchblende. The probability that a radioactive nucleus will decay is not influenced by temperatures, pressure, chemical process, etc., making it ideal as a tool for measuring the age of rocks that undergo extremes of all these properties. On the basis of radio-decay measurements on meteorites, on rocks returned from the moon, and on the oldest rocks found on earth, it is estimated that the solar system and earth are $4.5(10)^9$ years old and that the earth's present crust formed $3.5(10)^9$ years ago.

There are several other radio-decay processes that can also be used to date rocks. Among these is the decay of potassium to argon: $K^{40}$ (note that the common isotope of potassium is $K^{39}$) to $A^{40}$, with a half-life of $1.3(10)^9$ years. Because argon is inert it does not bond chemically to any other material and since it is a gas it can be assumed that it will be completely driven out of molten rock. Therefore, any accumulation must date from the time of solidification of the rock. Also to be mentioned is the change from $U^{235}$ to $Pb^{207}$, with a half life of $7.13(10)^8$ years.

## A Surprising Relation

We have already pointed out that the geologic eras, periods, etc. are defined by the life-forms found within their rock layers. The beginnings and ends of these intervals therefore mark distinctive changes in the development of life on the earth. Now that geologic times have been accurately measured by radio isotope techniques, we can assign dates to the initiation of these eras,[8] as shown Table 1-4. Here we have the same intervals and events as we discussed earlier. In referring to ages, here and throughout the rest of this book, we use a variant of the paleontologist and archaeologist notation: "BP" for years Before the Present. Note that as we approach the present it is necessary to change from eras to ages, though always taking the major interval consistent with age in question.

A curious relation can be noted, among the dates on Table 1-4. If we construct a semilogarithmic graph showing the logarithms of the times for the initiation of each era, in sequence, against equal interval markers for the event numbers, we find the straight line shown Figure 1-3. This use of equal intervals is presumptive at this points; later it will shown that a particular mode of evolution, called "emergence" frequently results in such semilog-linear behavior. Figure 1-3 will be interpreted as implying that major stages of life development have been cyclic, with the event markers being cycle numbers. Despite the facts that the geologic intervals are not universally accepted, are often defined in different ways, and that their initiations are not well defined, the straight line on Figure 1-3 seems

TABLE 1-4
APPROXIMATE DATES OF GEOLOGIC INTERVALS

| NUMBER | ERA | SYSTEM | EPOCH | AGE | INITIATION (BP) |
|---|---|---|---|---|---|
| I | Azoic | | | | |
| II | Archeozoic | | | | 3.5–4.5 $(10)^9$ |
| III | Proterozoic | | | | 2–2.5 $(10)^9$ |
| IV | Paleozoic | | | | 5.5–6 $(10)^8$ |
| V | Mesozoic | | | | 2–2.5 $(10)^8$ |
| VI | Cenozoic – | Paleogene | | | 6.5–7 $(10)^7$ |
| VII | | Neogene – | Miocene | | 2.5 $(10)^7$ |
| VIII | | | Pliocene | | 5 $(10)^6$ |
| IX | | Quaternary | | L. Pleistocene | 1.8 $(10)^6$ |
| X | | | | M. Pleistocene | 7.5 $(10)^5$ |
| XI | | | | U. Pleistocene | 1.25 $(10)^5$ |

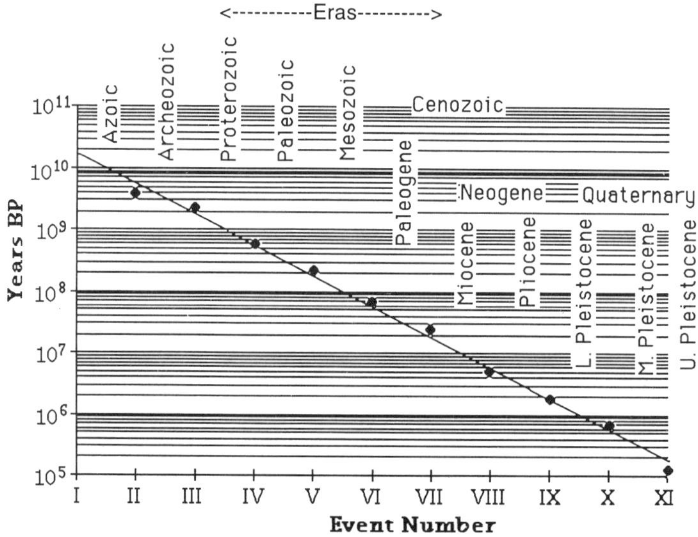


**Figure 1-3** Ages of geologic eras.

to imply that significant intervals of life development on earth, decrease logarithmically. We will discuss this at greater length in later chapters.

Extending this line to Event I brings it to the time of the grossly non-biological event of the birth of our entire universe — the "Big Bang." This is an amazing coincidence, if it is, indeed, a coincidence. It will be argued later that it may not be, and that its occurrence is consistent with the interpretation of general version of this figure.

Whenever we examine a graph of measured data, such as Figure 1-3, we should be aware that that there are possible deviations from the relation it purports to present, so that the points may not fall directly on the line. In the case at hand these can be mistakes in measuring extremely small quantities, e.g., the amount of radioactive material in a rock sample, or inaccurate decay rates of the radio emission process, or the changes may not have occurred uniformly at all points on the earth. As a result the dates are uncertain and the excellence of any linear relation between them is, to some extent, fortuitous. It is the overall conformance of all the points, to the line, that supports the linear relation rather than the goodness of fit of individual points.

## Alternative Measurements

The radio-dating methods described thus far establish the ages of minerals in the rocks and therefore indirectly date the fossils contained in them. Furthermore, because of their very long half-lives, these radio-decay techniques are well suite for times greater then $10^6$ BP. For more recent events, e.g., only dating from the latter part of the Quaternary or early Holocene Epochs, a finer time scale is needed. For this, great use has been made of the carbon-nitrogen process, which has a shorter decay time and the advantage that it dates the fossils themselves. This cycle is initiated by cosmic rays, which are high energy particles that reach the earth from space. They collide with atoms in the atmosphere, breaking them into less energetic particles. A neutron from these interactions can combine with one of the abundant atmospheric Nitrogen ($N^{14}$) atoms which then becomes unstable and emits a proton to become carbon ($C^{14}$). The $C^{14}$ nucleus is also unstable and decays with the emission of an electron, returning to $N^{14}$, with a half life of about 5600 years. The result of this continuous, dynamics process is an equilibrium concentration in the atmosphere of one atom of $C^{14}$ for $10^{12}$ atoms of the stable isotope $C^{12}$. The most common location of carbon atoms is in atmospheric carbon dioxide that is absorbed by living things where its relative isotopic concentration is also $1{:}10^{12}$. Upon death the uptake of carbon ceases, but decay of $C^{14}$

continues so its present concentration in fossils yields an estimate of the elapsed time since death.

Two problems beset the $C^{14}$ method. A small inclusion of debris from the fossil excavation or dust from the analysis laboratory can seriously affect the dates obtained, particularly when dealing with ages near the limits of the technique, 40,000 BP. In addition, beyond 10,000 BP fossil bones loose collagen, and with it the carbon atoms being tested, and they tend to absorb carbon from the environment. Thus, although great improvement have been made recently, this dating method becomes unreliable beyond approximately 30,000 BP. On other hand, the K-Ar determinations are useful only beyond 300,000 BP. The intervening interval constitutes a major period of evolution of the human species so that establishing our own family tree demands other ways of assessing fossil age.

This has long been a serious archaeological lacking but several physical techniques have now been proposed to fill this gap. Prominent among them are the methods of electron-spin resonance (ESR) and thermoluminescence (TL). Both of these measure the fossil sample itself rather than the surrounding rock and both derive from standard techniques for testing the structure and properties of solid-state electronic materials used by engineers and physicist in the microelectronics industry. Their basic premise is that imperfections and faults in the atomic arrangement of solid materials act as traps to hold some of the electrons of that material. In the case of fossil samples, these faults are induced by interactions with high energy radiation such as cosmic rays from outer space or alpha and beta particles from terrestrial radioactive minerals. The number of trapping centers therefore depends on the elapsed time after death that the fossil has been subject to these sources. If the rate of trap creation has remained constant during the period of interest then ESR and TL can measure the number of traps and establish the time since death.

In the ESR method a steady magnetic field is applied to the sample being studied, causing alignment of the magnetic moments of its electrons. A radio frequency electromagnetic field is then also applied, causing the moments to oscillate (or resonate) about the fixed field. By measuring the strength of oscillation, i.e., its resonance amplitude, the number of participating electrons can be determined. Because the electron can be regarded as a spinning top whose spin and magnetic moment are related, this phenomenon is called electron-spin resonance. One advantage of ESR lies in the fact that the electrons are examined but not changed by the measurement so that many tests can be made on the same small sample.

In TL testing the sample is carefully heated. At some point the thermal energy entering a trap causes it to release the electron, surrendering its energy as a photon of light. By measuring the strength of this luminescence, the trap density can be determined, and is proportional to the sample age. Unfortunately, once the electron is released the trap is empty so the test cannot be repeated.

While both TL and ESR cover time spans to $10^6$ BP, they are also subject to sources of systematic error that need compensation. For example, some investigators have questioned whether the radiation dose rate, creating the electron traps, was truly constant during the tens and hundreds of thousands of years involved. Another major difficulty arises because the water solubility of uranium allows this mineral to move through the environment. As a result it may be absorbed by buried bones and teeth where its radiation products alter the density of electron traps. Investigators have learned to correct for these confounding effects, depending on local and overall fossil location, but doubts remain and some differences exist between the various techniques for age determination.

## Chemical Dating

Means have also been proposed to establish dates from protein molecules that are trapped in some fossils but these have not yet achieved general acceptability. However, one chemical technique is worth noting because, although it does not establish the age of fossils, it has been used to study the specific descent lineage of mankind.

It will be pointed out in a later chapter that the energy sources in the cells of most living things are their mitochondria, small bodies that store available energy in the easily released chemical bonds of the molecule adenosine triphosphate (ATP). These mitochondria are unusual in at least two respects. At the time of conception, they are found in the large ovum contributed by the mother, and not the fertilizing sperm contributed by the father. Also, mitochondria contain their own DNA, deoxyribonucleic acid. DNA is predominantly found the chromosomes of the cell nucleus that carry inheritance. However, when cell reproduction occurs this mitochondrial DNA (mtDNA) also replicates itself into each daughter cell. As a result, it constitutes an inheritance that is handed down exclusively through the maternal line.

Since mitochondria are found in all nucleated cells, they must be among the oldest adaptations of life. Because they are so universal and relatively invariant across broad family lines, it is assumed that they are not affected by the normal pressures to change that are found in evolutionary development. Any changes in the mtDNA are therefore due solely to the incidence of mutagenic radiation, i.e., gamma or other high energy emissions that cause changes in the DNA. These mutations are measured to be between 2% and 4% per million years. While there are arguments against assuming a uniform mutation rate of the mtDNA, it is generally accepted. mtDNA differences can then be taken to be proportional to elapsed time and constitute a means to trace change.

By an elaborate process of physical and chemical separation, using the ultracentrifuge, pH sensitivity, and chemical solubility differences, the mtDNA can be isolated. A process of "electrophoresis" is then employed, in which an electric current forces the DNA molecules to migrate through a gelatinous substrate. The molecular mobility in this process depends, among other things, on its detailed molecular structure, and therefore on the number of mutations it has experienced. As a result, differences of molecular structure translate into differences of migration distance, giving a spatial display that shows the molecular differences and, therefore, the mutation changes between different samples.

Several studies have been reported[9] that take samples from members of different groups from around the word. These include African aborigines; Australian Bushmen; north, central, south, and coastal native Africans, Orientals and Caucasians from different regions, Polynesians, North and South American Indians, etc., not to mention various monkeys and great apes. From their different mutations, investigators try to construct relational maps and assign dates at which various groups became separated from a common ancestor. There is a great deal of debate about how to do this, with different analytic programs frequently leading to conflicting results. Some of these results have been related with the development of racial differences, and some related to the development of language distinctions.[10] Some results are striking, such as that the interracial differences between humans are much smaller than the separation from our nearest living primate cousins, and that variations within each human race are greater than the mean differences between racial groups. From these results it appears that we are, essentially, one genetic, human family. Because this method traces matrilineal descent, one of its publicized results is the conjecture of a primordial Eve, i.e., a single female or a small group, from whom all mtDNA lines emanate.[7] Considerable objection has been raised to this particular result[11] and, as stated above, these studies are generally controversial. However, we note that they do seem to support what has been called the "out of Africa" scenario of human descent, to be discussed later.

This chapter has dealt with the measurement and mode of expression of geologic times. With this background, we now enter into a description of evolution itself in the next three chapters. From that discussion, we will be able to extract certain critical features that conform to a coherent, cybernetic explanation of the evolutionary process.

## Endnote

[A] While we will not use them in this book, the logarithms of number less than one present no difficulty if we recall the use of negative powers of ten, e.g., $\log 10^{-4} = \log 0.0001 = -4$. Thus, the number 0.0003162 has a logarithm of $-3.5$ since $0.0003162 = 0.0001 \times 3.162 = 10^{-4} \times 10^{0.5} = 10^{-4+0.5} = 10^{-3.5}$. Numbers between 0 and 1 have negative logarithms.

# 2: GENESIS

## In The Beginning

> In the beginning God created the heaven and the earth.
>
> *Genesis 1:1*

> There came into being from the [thought] and the [speech] the form of Atum. The mighty Great One is Ptah who transmitted life to [the creator-god Atum] ... Thus it happened that it was said of Ptah: "He who made all and brought the gods into being."...
>
> *Creation myth from the First Egyptian Dynasty*[1]

All peoples have pondered the very utmost beginning, the creation of all the world. Creation myths served to rationalize the order of things in terms that could be understood at their time and, within their ability to comprehend, to express philosophies about man and his place in the world. Explanations in the modern "objective" sense were hardly sought. In making a comparison of ancient and recent beliefs, it is worth making note of two specific features.

Even stories of creation must have a beginning. In the quotations above, the initial point assumes, without comment, that God or Ptah is already present. In this regard we should note that in early Indo-European languages the words for "begin" and "beginning" are associated with the ideas of "entering into" or "becoming part of"a process (that is underway).[2] As a result, no earlier explanation was necessary.

The view of creation represented by these quotations deals with the beginning of "things" but does not deal with the beginning of time. While modern cosmological theory has addressed this distinction, its response interrelates them in a way that raises other issues. We will discuss this briefly in a later paragraph.

Both myths describe a process in which the deity creates the world from nothing-*creatio ex nihilo*, and the descriptions lack detail; being so far from experience that none could be supplied and, in a strictly teleological sense, none was called for. In contrast, not only do present day physicists and mathematicians conjecture about the initial state but they have established a detailed sequence of the events that immediately followed an instant of cosmic creation. Their descriptions result from mathematical extension and extrapolation from physical theories of the behavior of the known universe, including nuclear, quantum mechanical, relativistic, gravitational, electromagnetic, and cosmic properties. For the moment, let us start with the statement that between 10 and 20 billion years ago [i.e., 1.0–2.0 $(10)^{10}$ BP] a unique event occurred, the "Big Bang" that spawned the present cosmos.[3] In spite of the seeming similarity, it is a mistake to regard this event as a situation of *creatio ex nihilo* since the preexisting state already had some specific characteristics.

An analogy to this event can be drawn from two other concepts that are found in fundamental quantum physics. One is the vacuum ground state. To most of us, vacuum is a region devoid of the matter and absolutely unchanging. However, in the quantum mechanical description of such space the vacuum is not quiet. Rather, it experiences random fluctuations of energy, both positive and negative, from its rest or "ground" state. Associated with this is the second concept, the uncertainty principle, specifying that energy fluctuations can exist for very short periods of time, determined by the product relation: $\Delta E \Delta t < h$. Here the "delta" symbol, $\Delta$, denotes a change, so that $\Delta E$ is the change of energy and $\Delta t$ is the duration of that energy change. h is Plank's constant, which is the fundamental scale factor of the quantum mechanics of atomic and high energy systems. It is very small: $h = 6.62620(10)^{-34}$ Joule-sec. The Uncertainty relation tell us that spontaneous energy changes can occur but that high energy changes relax back in extremely short times.

If the energy change is large enough, it can result in the creation of mass. From Einstein's equation $E = mc^2$, mass, m, and energy, E, are equivalent, though a great deal of energy is needed to produce a small mass because the factor $c = 3(10)^8$ m/sec, the velocity of light, is so large. This relation is in conflict with the classic laws of conservation of mass and of conservation of energy, which say, respectively, that mass cannot be created or destroyed and that energy cannot be created or destroyed. Einstein's relation means that these laws must be combined into a single law of conservation of mass-energy that applies at high energies. Therefore, within the uncertainty relation and without violating this new conservation law, a negative and a positive charge, each with some mass, can be created spontaneously. (Note that there is also a law that charge must be conserved so we need the equal and opposite charge.) These particles would be expected immediately to coalesce back to the ground state. However, if the charged masses should be created with sufficient kinetic energy that they separate appreciably in the allowed time interval, then they could no longer recombine and particle creation would have occurred.

An analogous event, on a much, much grander scale, is postulated to have occurred at the beginning of our universe.[4] After this very brief period of creation (about $10^{-35}$ sec.), called inflation, an explosive expansion called the Big Bang took place. We note in passing that the energy involved in this event need not have been as great as we regard the energy of the visible universe. In their energy bookkeeping, physicists attribute negative energy to the gravitational concentration of matter into stars and galaxies. As a result the total cosmic energy "bottom line" may not, itself, be of cosmic size.

For some time after the Big Bang, the universe was fantastically small, dense and hot, and the physical laws governing its behavior were very different from those with which we are familiar. The series of events of its evolution must be inferred from allowed solutions of the quantum mechanical and relativistic dynamics. This mathematical description is still being developed, and there is considerable discussion about features of the theory. These include the nature of the state from which the inflation developed and the existence and overlap of other universes that may have been created the same way.[5]

In Newtonian mechanics, the basic dimensions are length and time. Fundamental properties of matter, such as mass and electric charge, are described in terms of these dimensions through the laws of force each displays — the gravitational law for mass and Coulomb's law for charge. In contrast to this, Einstein's Theory of Relativity has the speed of light as its basic measure, with distance and time as derived quantities, and considerable attention is paid to the way in which they are measured. Because length and mass are strongly interrelated, the distance and time between events are not absolute but depend upon the presence of mass. Time is a derived property and in the absence of mass it cannot be defined. This is the basis of the statement that before the Big Bang, i.e., before there was mass, there could be no time, so that $t = 0$ at the Big Bang.

In any event, the description is detailed and leads to some observed features of our universe. Thus, at some point in the evolving state, perhaps a few hundred thousand years after its initiation, its temperature became low enough that particles "precipitated" out of the energy system and electromagnetic radiation became distinct. This is analogous to the way a chemical reaction might precipitate solute particles, leaving a clear solvent. That primordial radiation uniformly pervades the entire universe today. As predicted, it is now measured[6] to be a much cooled afterglow; its characteristics are those of the radiation from a body whose temperature is only 2.734 K (when speaking of absolute temperatures it is customary to omit the notation "°" for degrees Kelvin). Zero Kelvin corresponds to –273.16°C.

For several years after the discovery of this background radiation, its apparently extreme uniformity was a source of difficulty for astrophysicists. Because it represents the state of that very early universe, its unvarying property, regardless of the direction from which it is measured, indicated that the cosmos itself was equally unvarying, being indistinguishable from point to point. The difficulty arises by comparing that early uniformity with the granularity of the present universe, as manifest in the grouping of matter into stars, of stars into galaxies, of galaxies into galactic groups, and of these into "walls" and planes on a grand cosmic scale. It raises the question of how the present coarseness and structure could arise from a condition that is completely without variation. Then, early in 1992, observers announced[7] that measurements made with great care and precision, using satellite observing facilities had, indeed, indicated that there were fine fluctuations in the observed background radiation, amounting to about 10 parts out of $10^6$. This is small, but just big enough that it would be expected to have led to the present condition.

The background radiation is not the only supporting evidence for the Big Bang theory. It appears that the momentum of expansion of the Big Bang is still with us since the contents of space, the galaxies, are moving apart like raisins in an expanding cake dough. This is verified through the Doppler Effect. To understand this effect we must know that for wave motion, such as of light or sound or radio waves, the relations holds that $f = c/\lambda$, i.e., the frequency of the wave, f, equals its velocity, c, divided by its wavelength, $\lambda$. For light, c is a universal, unvarying constant. Figure 2-1a shows a single wave pulse being emitted by a stationary source, on the right, and received by a stationary detector at the left. In Figure 2-1b the emitter is moving away from the receiver during the emission process. Because emission takes a finite (though very small) time, the wave has its initial point of emission closer to the receiver than its final emission point. As a result the length of the wave is increased relative to the wave from the stationary source. As $f = c/\lambda$, frequency is inversely proportional to wavelength, so that the longer wavelength reaching the receiver will be interpreted as coming from a source with a lower frequency. For light, the received signal seems to be shifted toward the red (i.e., long wavelength) end of the visible spectrum; for sound waves the shift is toward lower pitch.

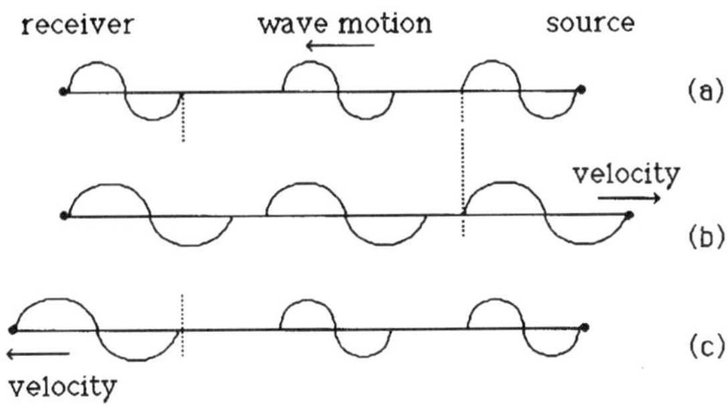


**Figure 2-1** The Doppler effect.

Figure 2-1c shows a stationary source and a detector that is moving away. Now the travelling wave must catch up with the detector so that the time between its initial and final interactions, for one cycle, is greater than if the receiver is not moving. As a result of the receiver motion the wave period, T, seems longer. The frequency of a wave is the inverse of the period ($f = 1/T$) so it is interpreted as a lower frequency. These behaviors describe two aspects of the Doppler Effect, the shift toward longer wavelengths, or toward lower frequencies, of waves transmitted between an emitter and a detector that are moving apart. The change of visible light toward the red end of the spectrum gives the phenomena its popular name: the "red shift." Of course there is an opposite Doppler shift for objects moving toward each other, which might be called a "blue shift." The corresponding change is observed with sound, where an approaching siren has a higher pitch, but only a red shift is found astronomically, indicating that cosmic bodies are moving apart.

In the 1920's Edwin Hubble studied certain stars, the Cepheid variables, whose brightness changes periodically; these objects have an invariant relation between their average brightness and its frequency of change. Hubble used this property together with the Doppler shift to study Cepheids in our own "neighborhood" as well as in distant galaxies and galactic groups. He found that more distant systems are moving away from us at greater speeds than nearby ones. He measured the proportionality constant between distance and velocity of recession, called the Hubble constant. Not only is Hubble's relation consistent with the concept of an expanding universe, but the Hubble constant, relating distance and the degree of red shift has become the standard method for determining the distances of cosmic systems.[8] Distances in the cosmos are so great that they are measured in "light-years," the distance traversed by light in one year while travelling at $3(10)^8$ m/sec (or 186,000 mi./sec). From their red shift, we can conclude that our instruments detect objects that are thousands of millions light years away. This means that the radiation we examine was emitted that long ago. The most distant visible objects in our expanding universe are so far away that light takes $1.5(10)^9$ years to reach us; in those cases we are looking at an early period of the universe.[9]

Recently some astrophysicists have reevaluated Hubble's constant, based on theoretical expectations of the properties of certain supernova that are observed in distant galaxies, and have arrived at different values, depending on the theoretical model employed. Their values indicate that the universe may be only half as old as indicated above. This results in the awkward situation where the ages of the most distant galaxies appear to be less than the ages of some of the stars they contain. This disagreement need not affect our present description, but we should note that it is representative of the exciting, dynamic state of astrophysics as it seeks to understand the cosmos in which we find ourselves.

The atoms and molecules that emerged following the Big Bang were eventually drawn together by gravity, within relatively local though still cosmic distances. This led to the formation of galaxies more than $10^{10}$ BP, of our sun approximately $5(10)^9$ BP, and of the Earth about $4.6(10)^9$ BP. The energy released by its gravitational condensation heated this new planet, causing it to be in an initial molten state until it cooled by radiation into space. As a result, the formation of a crust did not occur for another $10^9$ years. Even with its crust of light basaltic minerals floating on the viscous mantle, the early Earth was, if anything, inhospitable. Its surface was chaotic and changing and under intense bombardment by meteoric debris from the solar system condensation, and its atmosphere was composed of noxious, volcanic gasses which are inimical to life as we most familiar with it.[10, 11]

We should stop here to catch our breath. The last paragraph has raced through the early history of the universe up to the cooling of the Earth from about 15 or $18(10)^9$ BP to $3.6(10)^9$ BP. Our use of scientific notation tends to hide the significance of these time intervals; note from the exponents that they are in units of 1 000 000 000 years. The times involved are beyond imagination — the total amounts to 80% of the age of the universe, and that is only to arrive at the point where events on the earth can begin.

## The Origin of Life

One of the basic features of life is the ability to reproduce itself. Although other processes may appear to present a form of reproduction, such as crystal growth from an initial "seed" crystal, these are more replication processes than reproduction in the biological sense. Distinguishing features of life-form reproduction are the complexity and variability of the organic generative process and products and, despite these, the overall uniformity of those products. In each life-form, the master plan for this process is written in its genes in the chromosomes that are generally within the nucleus of its basic cellular structure. These chromosomes are very long chains of DNA — deoxyribonucleic acid-the famous "double helix" of intertwined molecules. Because each helix is comprised of complementary paired chains, it can separate into its two equivalent parts, allowing each to reproduce its counterpart from the enzymes and other materials in the cell. This replication is what occurs during reproduction and, as a result, each daughter cell resulting from the division of a parent cell carries an identical master plan. In this way a duplicate of the parent is produced from cell to cell, then from one generation of a life-form to the next. The mechanism of realizing this blueprint, that is of building the living structure it encodes, is through RNA-ribonucleic acid. RNA molecules copy the code and translate it into a series of amino acids, the proteins that make up the enzymes and body of each cell.

Nucleic acids, the NA in DNA and RNA, carry the information for their own replication, for building cell and body structures, and for synthesis of the enzymes that mediate cell functions. It follows that DNA would have had to precede the formation of the first cell. Yet the cell protein enzymes and catalysts are necessary for the production of that DNA. Since neither could precede the other, a problem exists in understanding how the true origin of life could have occurred. DNA, RNA, and the other materials that constitute living matter are called "organic" materials. They are very long chains of hydrocarbons, being built on a base of carbon atoms, primarily with hydrogen, nitrogen, and oxygen. On its most fundamental level, the question of the beginning of life reduces to that of the origin of the first replicating, organic molecules.

Several proposals have been made to resolve this dilemma, some of which can be readily discarded. One suggests panspermia, that is, that an initiating, simple life form or precursor arrived on earth from a distant solar system. Not only does this lack any substantiation or method, but it does not really solve the problem of origin; it merely moves the location of its resolution from earth to some other site in the galaxy. We also ignore the unfounded proposal of "spontaneous generation," that a vital force of complex life forms sprang into being spontaneously and complete. Similarly, it has been suggested that, when conditions were "right" a unique event occurred, or at least an extremely improbable one, that created the necessary reproducing organic molecules to initiate life. These proposals are counter to our concept of "process" whereby the whole of evolution, from its earliest moments, is a continuous series of developments, each being reasonable, comprehensible, and likely in its own milieu.

The acceptable, more useful, and penetrating models can be divided into those specifying extraterrestrial and those invoking terrestrial origins. The differences are partly based on uncertainties about the atmospheric composition of the newly formed, volcanic planet.

Extraterrestrial scenarios[12] originate with the observation that a certain type of meteorites, called the carbonaceous chondrite meteorites, and their presumed parent bodies, the C type asteroids, are 3 to 5% rich in complex organic materials, and some have considerably more. About 3% of the carbon in these bodies is in amino acids, the basic building blocks for the protein enzymes and catalysts. This must have formed in the interstellar cloud from which the sun and Earth condensed. At the time of this condensation and subsequent cooling of the Earth, there was also a great density of these asteroids and of cosmic dust, which rained on the Earth. If the atmosphere of early Earth consisted of nitrogen and carbon-dioxide, at 10 to 20 times the density of our present atmosphere, these space rocks would be slowed enough to not be destroyed by atmospheric friction and heating as they fell to earth. They could successfully deliver a considerable amount of prebiotic material to the surface.

In considering terrestrial models, it is assumed[13] that the atmosphere of the primitive Earth was made up of primarily methane and ammonia, as well as considerable amounts of carbon-dioxide. In this event, there is evidence that some organic amino acids could have been produced by the action of high energy ultraviolet radiation on hydrocarbon molecules in the early seas and lakes, or by quenching of high temperature atmospheric reactions created by volcanic shock. It has also been suggested[14] that silicate molecules, which form long-chain molecular bonds resembling those of carbon and which are present in clays and muds, could have acted as templates for formation of the first organic chain molecules such as RNA.

Recently much attention has been paid to the proposal that RNA alone acted as the early template and translator and that only later, when protein was more abundant, was DNA co-opted as a more efficient agent.[15] This scenario has found support in the observations that RNA can serve as more than a mere transcription messenger in the cell. Recent experiments have shown that RNA can also serve as a separating, joining, and bonding agent for the protein bases, giving it the necessary functional properties.

In any event, by $3.4(10)^9$ BP bacteria had appeared in the lakes and oceans.[10,16] These are described as having procaryotic cells, that is, collections of simple organic compounds within a fatty membrane, with no nucleus, and which reproduce by simple fission after copying their single looped DNA chain. The lipid envelope isolates the cell interior so that only selective chemical exchange can take place with the exterior and the interior chemical environment can be enriched and optimized. Early procaryotic cells probably obtained energy from spontaneous chemical processes, such as glycolosis, which release energy originally stored by other processes, e.g., in the form of glucose, using intense solar radiation.

About $3(10)^9$ BP, the blue-green algae or "cyanobacteria" appeared. These are able to absorb solar energy directly and, through a series of steps, to use the inexhaustible supplies of carbon dioxide and nitrogen to store that energy in more usable form in simple sugars. At first this process may have depended on the noxious gas hydrogen-disulfide as a needed electron donor for the chemical process, but later, when water became freed from its form in the rock substrate, it was a more available source. The blue-green algae were remarkably successful and they spread throughout the oceans of the world. Their photosynthesis process released oxygen into the atmosphere and, over the course of a billion years, the accumulated oxygen caused tremendous changes.[10]

Among these was the generation of high altitude ozone (a highly reactive form of oxygen) the shielded the earth's surface from the deadly high levels of solar ultraviolet radiation. This allowed bacteria to leave the ocean depths and approach the surface. However, as oxygen is highly reactive its concentration in the new atmosphere and dissolved in the water was toxic to them, and they had to develop ways of dealing with it. This may have led to the formation of eucaryotic cells and to oxygen mediating enzymes. Respiration in these new cells involves a chain of processes, mediated by chemical catalysts and enzymes, and resulting in the combination of oxygen and sugars. This process results in the release of much greater amounts of energy than was previously available, and its development opened the way for the next stage of life.

Colonies of procaryotes had existed but with the appearance of eucaryotes, about $10^9$ BP, joint existence and symbiosis took on a new meaning. The eucaryotic cell, from which all subsequent life evolves, is some 10 times larger than the procaryotic and its volume is therefore 1000 times greater. Each of the internal components, such as the nucleus and the energy-storing mitochondria, is enclosed in lipid membranes. This arrangement has features of a colony of bacteria that have banded together. L. Margulis[17] has suggested that this configuration protects them from the hostile environment and allows them to function symbiotically. It may have originated when one form engulfed another and retained it internally when it was found to be beneficial, a process called endosymbiosis. Within their cloistered environment each component reciprocally supplies the needs of the others. Over the course of time the individual members have surrendered most of their genetic prerogatives to the central nucleus, which has 1000 times the DNA of a procaryotic cell. In considering this description, it is important to recognize that, in fact, "The origin of the defining feature of eucaryotic cells, the nucleus, is not understood".[18]

Not all eucaryotic DNA resides in the nucleus. The cellular structures called mitochondria contain much of their own DNA in a configuration very much like that of the earlier procaryotes. Mitochondria store high concentrations of readily available energy in molecular adenosine triphosphate (ATP). They move about the cell, surrendering their store and enabling the chemical process of life. Being common to all eucaryotes, mitochondria probably represent one of the earliest evolutionary developments. In Chapter 1 we indicated that the study of mitochondrial DNA is one approach to establishing recent (geologically speaking) human lineage.

Beyond the advantages of the eucaryotic structure, colonies of eucaryotic cells could be even more robust, and over a period of 100 million years yet a different sort of symbiosis was developed. It was to the colonies' advantage for different parts to serve specialized functions. For example, stability was improved if a region near its base developed an anchoring mechanism to the underlying floor. And the gathering of needed chemicals and proteins from the surroundings was likely improved if the uppermost cells grew wavelike hairs, cilia, to move the water more efficiently into a cavity. Within the cavity they could be metabolized by specialized cells with more efficient enzymes for this purpose. The chemical messengers and physical influences producing these changes in different parts of the colony became part of the reaction between and within the cells. Each part had become a specific type of tissue of the organism in which they functioned. The tissue members of those colonies that were most successful passed their abilities to specialize to their progeny, causing the colonies' common genes to have different expressions in different parts. The genetic and chemical coordination was so tight and mutual dependence so strong that the colony lived, reproduced, and died as a unit; the variously functioning parts has become organs of a single being.

It has only been in the last 40 to 50 years that microscopic examination of geologic deposits from those early periods have revealed the history of procaryotic and early eucaryotic cells and tissues. However, with the development of specialized tissues and organs, the ability to capitalize on varied environments and circumstances became almost limitless. There was an explosion of life and of life-forms around the earth that is readily visible in the fossil evidence. Each type differentiated as needed to fit the multitude of ecological situations in which it found itself. This rapid radiation of life forms occurred during the early Cambrian period of the Paleozoic Era, starting approximately $0.5–0.6(10)^9$ BP.[19]

Studies have revealed the existence of simple eucaryotic life well before the Cambrian onset, but without the variety and quantitative leap of that transition. It has been suggested that the critical feature leading to the Cambrian expansion was the appearance of sexual reproduction.[20] In the next chapter, we discuss the fact that this greatly enhances the opportunity for mixing of inherited characteristics and for generating a wide variety of structural and behavioral types. Some of these variations had the specializations mentioned above to enter different ecological positions and, in fact, to create new environments. The early Cambrian fossils disclose a tremendous variety of life forms, as though a system of trial-and-error was prevalent.[21] Many, if not most, were unsuccessful and did not survive. What remained after a relatively short period formed the basis of present life on Earth.

# 3: BIOLOGICAL EVOLUTION

## The Evolutionary Paradigm

Alfred Russel Wallace and Charles Darwin published their ideas on evolution in concurrent articles in 1858, and Darwin's great book, *The Origin of Species* appeared in 1859, greatly expanding and interpreting the supporting evidence. The centerpiece of their theory is "natural selection", i.e., the survival and propagation of individuals and species that are best able to endure interspecies and intraspecies conflicts, to survive predators, to compete for space and for food, to combat disease, to endure climatic and weather effects, etc. This is the complex of factors that are referred to as "the environment". Darwin proposed that animal and plant variants continually and randomly appear, and that the natural selection process of the environment allows the most well adapted and those with appropriate reproductive strategies to bear viable progeny, passing on their advantageous traits. During the course of hundreds and thousands of years these variations accumulate, producing inherited changes of morphology and behavior. Eventually a new species results through common breeding of these individuals, while older species might become extinct as their traits are found lacking under changing circumstances. This process describes a continuous, slow process of change.

Darwin's theory is neutral concerning the source of the variations. Mechanisms of inheritance were unknown at the time, although most naturalists held to the ideas of Jean Baptiste de Lamarck. Fifty years before Darwin's work, Lamarck had proposed that creatures pass to their progeny those characteristics acquired during their own lives. To quote Darwin:

> Lamarck was the first man whose conclusions on the subject excited much attention. This justly celebrated naturalist first published his views in 1801 ... In these works he upholds the doctrine that all species, including man, are descended from other species. ... With respect to the means of modification, he attributed something to the direct action of the physical conditions of life, something to the crossing of already existing forms, and much to use and disuse, that is, to the effects of habit.

Lamarck's belief was one of blending inheritance of the characteristics contributed by each parent. This led to difficulty in species differentiation since individual changes would be diluted by blending throughout the entire mating population.

It wasn't until 1900 that the work of Gregor Johann Mendel was rediscovered. Mendel was an Austrian botanist and monk who studied the reproduction of characteristics in the pea plant. In 1866 he discovered that heritable characteristics occur in pairs, carried by the germ cell, and having the properties of dominance or recessivity. By 1900, it had been recognized that acquired characteristics are never inherited, ending Lamarckian dominance, and Mendel's genetics was a timely and needed mechanism of inheritance. A new, Neo-Darwinian understanding, a "Modern Synthesis" of cause and effect, was forged in the early decades of the twentieth century.

The chromosomes in the nuclei of most cells occur in pairs and such cells are said to be "diploid". During mitosis, the normal process of cell division by which living things develop, grow, and repair themselves, the chromosomes of the parent cell are reproduced, giving each daughter cell the complete, double set of this genetic information. In the special case of creating germ cells for sexual reproduction, i.e., sperm and ova, the paired chromosomes are not duplicated, so that each offspring cell has only a single set of chromosomes. This process is called meiosis and the product cells are said to be "haploid".

In either case the division process is not simple. Among other activities that occur are crossing over, intertwining, and sliding of the complementary chromosomes before separation. During this motion, the chromosomes exchange material in large and small segments. Normally this does not alter their being equivalent to their original selves but, occasionally, errors and changes occur that produce new gene sequences.

These can be the result of deletion of one or more sections, insertion of new or duplicate sections, or inversions of the order of components. It is estimated that between 10 and 100 errors occur in each cell division.[1] In addition, one might also have molecular change of a gene, induced by high energy radiation, such as from radioactivity within the earth or from cosmic sources. Such radiation can alter, disrupt, or destroy the chromosome. All these errors and changes are called mutations.[2]

The complete set of genetic instructions of an organism is called its genotype. When those instructions are fully executed the physical structure that is generated — the resulting plant or animal — is called its phenotype. The appearance of mutations in the reproductive gamete means that the genotype resulting from the union of that haploid gamete and its haploid sexual counterpart, to generate a diploid zygote, is modified from its parents. Most mutations appear to have no effect on the phenotype; they are said to be silent, or synonymous. While most nonsynonymous changes are not viable, those that are provide the genetic variation for an altered phenotype that is acted on by environmental selection and is taken to be the source of evolutionary change.

Since the adaptability of species is generally limited to small modifications of their existing state, a long standing question concerns the rate of speciation, i.e., the time necessary to evolve new species. Following Darwin's own belief, it had long been held that the incremental genotype and phenotype changes of species accumulate during long geologic times until their total effect can be recognized as a new species. However, it has recently been acknowledged that this is not in accord with the fossil record which only infrequently shows the expected intermediate stages. In 1972, in response to this observation, N. Eldridge and S. J. Gould proposed a mechanism by which speciation can occur in short enough times that it is not likely to be manifest in the geological record.

According to their proposal,[3] which is quite widely accepted today, species exist in stasis, i.e., relatively unchanged, for very long periods. Genetic changes are diluted by matings in the larger, unmutated population. However, if it should happen that a relatively small number of members of the species are isolated and beçome subject to different environmental conditions, they can respond by changing. A new species can evolve and then spread beyond its original geographical locale.

It has generally been held that such change requires an "allopatric" condition, i.e., a physical separation of the parent and distinctive members, such as by the intervention of a mountain ridge, or isolation on an island. Recently, however it has been established,[4] by careful examination and the new techniques of molecular genetics, that speciation can occur even under sympatric conditions, where the two groups overlap in geographic domain. An example is a particular apple maggot which initially separated into groups that were attracted to different types of fruit trees. Through whatever mechanism they choose mates, in a very short time this was sufficient that an effective barrier appeared, leading to the evolution of non-interacting groups, despite their sympatry. Varied mechanisms can initiate the group division, ranging from separation by a river to the different, particular parasites to which the members are host.

In relatively short periods, frequently much less than the tens of thousands of years formerely expected, a new species can evolve and then spread beyond its original geographical locale. Because the long periods of species equilibrium are broken by these short periods of species change, this mode of evolution is called "punctuated equilibrium".

## The New Understanding

In the "Modern Synthesis" of evolutionary understanding, just described, chromosomal changes are attributed to chance events, and are random in nature and direction. It is similarly assumed that the environment is uncontrolled and impartial in its selection. This scenario therefore implies that the entire process is driven by chance and by forces external to the individual, who merely reacts to them as best it can.

Within the past 25 years the understanding of environmental influences has become vastly more sophisticated. It has been recognized that the environment to which a living thing reacts is largely constituted of and created by interactions with other creatures of its own and other kinds. At least in part, it is a self imposed constraint on the group level. In addition, with the advent of highly refined chemical and physical tools, an excruciatingly fine examination of genetic-chromosomal structure and of the mechanisms of genetic change is underway. The results of these studies have led to doubts about some basic assumptions of the Modern Synthesis, and we are witnessing the development of a "New Understanding".[5]

Studies of genetic structure have shown that genes are not simple, invariant molecular instructions, as envisaged by the followers of Mendel, but have significant structure and dynamics.[6] In particular, genes do not individually control specific aspects of development; they function in relation to gene groupings, both nearby and on other chromosomes. This alters their modes of action. A key mutation can conceivably have wide ranging interactions and exert a disproportional effect during embryonic orthogenesis, and might even initiate a cascade of secondary changes, leading to the rather rapid appearance of phenotype differences. The communication between gene groupings also makes them susceptible to management and alteration by cellular enzymes, whose actions, in turn, can be affected and controlled by sensing systems that bring relevant information from the environment to the DNA molecules. Campbell[6] notes:

> [The] new era of genetics is disclosing a remarkable new type of biological function. Some genetic structures do not adapt the organism to its environment. Instead, they have evolved to promote and direct the process of evolution. They function to enhance the capacity of the species to evolve.

This tendency is known as "Molecular Drive"; the structures to which he refers are DNA sections that control the timing and expression of the functional genes, causing different physical manifestations to appear or to be suppressed.[6] Some of these DNA sections, transposons, may move about the genome (the total genetic structure), and it is believed by some that they represent powerful means of introducing novelty into the evolutionary process.[7]

We have noted that nearly all eucaryotic chromosomes contain large numbers of synonymous sections; some investigators now believe that they represent past or future genetic experiments[8] or are sites of genetic folding to produce compatibilities and contacts between distant elements. It has also been suggested that those sections and regions that are duplications of others, result from an endogenous genetic "selection" process. The concept here is that there exists an intra-genetic survival competition wherein each mutation and genetic section competes for presentation within the genome. Those that gain dominance, and that are compatible with the rest of the genome, are further subject to the constraint that any change or stasis they develop must, at the least, be consistent with and nondetrimental to the resulting phenotype, in its own survival competition.

There is considerable debate today about the true role and significance of such genetic changes and competitions. However, the existence of active mutational processes has initiated a recognition that evolutionary change may not be merely a random response to external forces. Endogenous change may proceed more rapidly than change influenced only by exogenous influences. This would clarify the mechanism of punctuated equilibrium. Indeed, if endogenous processes are strong, they would shift the evolutionary question from the seeking a mechanism for rapid, punctuational change to seeking a justification for the usual, long-term species stasis.

A second aspect of the New Understanding that is being debated, concerns the hierarchical levels at which evolutionary changes occur. The discussion above indicates that particular differences and changes might be regarded as the results of lower level survival competition among genetic demes, i.e., coherent elements of structure or form. At the "creature" level we can understand that Darwinian selection acts on the individual, who strives to survive and reproduce. On both the genetic and individual plant or animal levels, the evolutionary imperative is to survive and reproduce, so that one's genetic inheritance is passed to the next generation. In this view, phenotype changes result from genetic level competition, and species change results from the successes or failures of its members. The saltational nature of punctuated equilibrium casts this process in a somewhat different light since it implies abrupt change at the species level. This relates to the fossil record, which is replete with cases where entire species or classes have suddenly appeared or become extinct. In such cases, the species or class also appears to act like an individual with regard to Darwinian selection.

This phenomenology indicates that species change may occur *in toto*, as though the individuals involved are tightly linked. Such an interpretation implies that evolutionary change can proceed, or at least appear to occur, at different cladistic hierarchical levels other than the individual.[4] This idea is opposed by some,[9] who point out that the "sudden" feature of species changes must be evaluated relative to the long geologic intervals through which they are examined. They claim that the gradual method of anagenesis is adequate to explain at least some species variation, consistent with known rates of genetic mutation and phenotype change.

Other proposals can also be offered in support of inherent, higher level evolution. The facts that most mutations are recessive, and that genetic realization is often through the action of multigene families, can be taken to imply that major changes may not be realized until the entire genome has been potentially activated to a sufficient degree. This demands that a population of interacting, mating individuals must all be on the verge of change at the same time; when some final, activating incident occurs the entire population can then convert. In addition, Mayr[10] has pointed out that a group can influence its own evolution because:

> ... changes of behavior very often function as pacemakers in evolution, by leading organisms into new niches or environments, which exert a new set of selection pressures and thus may lead to major evolutionary changes.

In addition, the appearance of altruism seems to support the idea of the species-as-an-evolutionary-unit. Altruism will be discussed in chapter 4; it refers to the fact that individuals will frequently sacrifice themselves for the good of their cohorts, be they herd, pack, tribe, or whatever. This is understood in terms of survival of the common genetic pool, rather than of the particular individual's genes, and it implies a greater coherence of evolutionary significance.

## Small and Great Beasts

The beginning of distinct tissues and organs and of sexual reproduction were mentioned in the previous chapter, leading to the explosion of visible life forms that initiated the Cambrian period. Just why this abrupt change took place is difficult to understand other than to state that evolutionary development had advanced to the stage that it could happen. Since much of organic evolutionary change also appears to be environmentally driven, we note that the Cambrian followed the greatest glaciation in earth's history. This is in addition to the increasing atmospheric oxygen level, the strongly shifting surface of the earth, and a period of greatly enhanced sedimentary mineral deposits.[11]

Feeding on matter brought in by cilia was followed by the development of arm-like protrusions around the mouth to intercept larger food samples; a rasping "tongue" was followed by cartilaginous jaws with biting power.[12] Mobility is an advantage in the dispersion of larvae and in the seeking of and the avoidance of being food, and this lead to creeping and, eventually, to fins and projective body motion. These developments resulted in much more efficient predation, strongly influencing the direction of subsequent evolution, which now had to include active defensive mechanisms as well as aggressive food gathering.

Such changes influenced and were influenced by the development of sensory structures. In addition to sensing contact there developed sensitivity to light (rudimentary vision), to vibration (rudimentary audition) and to the chemical environment (rudimentary olfaction). These primal abilities were, at first, poorly developed and poorly integrated. With improved sensors, perhaps associated with predation or safety, a coordinating nervous net was needed. At first this was very loose but gradually developed into a nerve cord with ganglia as coordination centers, and then a prime center, a "brain". Generally, "chordates" are elongated and are bilaterally symmetric for more effective motion, and have a higher degree of sensory structures.

We have considered only the development of animal characteristics but there was plant evolution as well, and the latter had its inevitable effect on the former. With the rise of fresh-water vegetation, animal life followed into this new environment. More powerful swimming was needed in strong water currents, probably resulting in the replacement of a cartilaginous structure by the bony skeletal structures of the vertebrates,[13] which appear about $4.5–4.3(10)^8$ BP. In addition, reproduction is a more serious affair in flowing water since gametes can no longer be simply shed into the water for fertilization and development. Successful fertilization now requires the development of some sort of courting and mating behavior. About this time fish evolved a circulatory system using hemoglobin for enriched cellular respiration.

It is conjectured that animal movement from aqueous to terrestrial environments was initiated by severe seasonal droughts during the Devonian and Carboniferous Periods of the Paleozoic era. Those vertebrates having accessory air pouches that could serve as temporary lungs, were able to leave their evaporating, stagnant pools and lakes in search of another body of water. Tetrapod limbs from paired steering fins served for the overland trip and, over millennia, along with the development of new neural mechanisms, evolved into limbs for land transportation by these amphibious animals. The development of lungs from air pouches and the development of limbs from opportunistically located and oriented fins are examples of what has been called anticipatory — or pre-adaptation. This does not imply that the need for a particular change was anticipated but rather that a structure that was already present could be used to meet a need that arose later. If the previous structures had not been present a particular line of evolutionary changes might have ceased or, more likely, it would have proceeded in a different direction.

True amphibians appeared about $3.7–3.3(10)^8$ BP. These are partly aquatic and partly terrestrial. To prevent desiccation their eggs need a wet environment so their tadpoles are aquatic and have gills whereas adults are generally air breathing. Amphibians were superseded by reptiles as the dominant life-form. Being land dwellers and possibly also in response to droughts, reptiles developed a reproductive technique in which fertilization is done internally and their eggs, when they appear, are protected by a substantial shell. Within this shell is the reptilian amnion, a membrane enclosing a liquid chamber that provides the missing aquatic environment for the developing embryo and a large yolk that provides nutrients.

In our discussion of the evolution of animal life, we are presenting a linear development sequence and are ignoring characteristics and parallel histories of the many other life groups that arose and changed or became extinct. We should understand that there was and is a great diversity of life forms filling the earth. Let us, therefore, make a side excursion to examine the groupings into which biologists divide life on earth.

Taxonomic divisions are made on the basis of similarities of structural anatomy and anatomical functions, although physiology may be considered as well. For example, we have mentioned the phylum Chordata whose members have a longitudinal nervous chord that links the tissues and senses to a coordinating brain. Generally, the chord lies in the dorsal side of the creature, and in the majority of chordates, it is enclosed in a segmented bone structure, the vertebra, which is part of its bony endoskeleton. However, in some primitive chordates, there is only a semi-rigid supporting "notochord" in place of the bony skeleton and in some, e.g., the sharks, the skeleton is cartilaginous. Each such difference defines a different subgroup and it takes considerable study to classify a new species, whether living or fossil, in its proper taxonomic place. The divisions according to different criteria do not always coincide, so that categorization may be a matter of examining the overall structure and functioning, to determine relatedness of different species. In recent years a pragmatic method has been developed to resolve some of these differences. "Cladistics" is the study of the relationship resulting from shared inherited characteristics. It involves examination and measurement of a number of features of a species and ordering them relative to the similar features of other species. Using computer comparisons of the number of common features and their degrees of disparity, an ordering of relative similarity and change can be made. From this a "cladogram" is derived with the most likely descent lineage and relationships between the species. Of course, where there are narrow differences and alternative classifications, the systematists may disagree on fine points of the taxonomic assignment.

Taxonomic assignment presents an even greater problem for paleontologists who wish to place the fossils they study into biological groups so that comparisons can be made and trends of development made visible.[14] This raises an interesting question of the exact relation between taxonomy and cladistics. Is the phylogenetic development of a creature, i.e., its particular cladistic history, a result of its ancestors having existed in and passed through all lower taxonomic structures, or is its taxonomy merely a grouping of properties we perceive but without necessary historical relevance? It is generally assumed that the phylogony of a species, parallels its taxonomy, although this may be questioned in specific cases.

The biologist's taxonomy groups depend, primarily, on body functioning and construction as exhibited by soft tissues, i.e., not solely on bones or shells or the lack thereof. Fossil remains of ancient creatures do not leave soft tissues so that classification requires considerable inference through understanding the relations between the positions and joinings of the fossilized remains and the functioning and necessary body aspects of the formerly living creature. Frequently, the hard fossils, or impressions left by soft parts, are meager indeed, leading to difficulties in deciding borderline cases between groups. Comparisons are made difficult, also, because fossil species are frequently found that have no living counterparts, with entire, great groups having vanished in times past.[11] However, in spite of occasional differences of paleontological and biological categories, there is good overall agreement. To facilitate our discussion we briefly review the divisions made.

Living creatures are commonly divided into two great Kingdoms: Plants and Animals, though the division is often made in three: Protista, Plantae, and Animalie, and sometimes in four: Protists, Fungi, Plants, and Animals. The Protists are single-celled, primitive organisms with intermediate forms that may be difficult to classify into the other kingdoms. These include bacteria, algae of all sorts, diatoms, fungi (in the three part categorization) and protozoa. Very recently it has been recognized[15] that, besides these, a very different lifeform, the Archaea, also exists on earth. These are microbes that live in extreme environments, such as hot volcanic flumes that rise from geologic faults in the ocean floor. It has been shown that they possess such a remarkably distinct genetic structure that they must be considered a different domain of life.

Plants constitute organisms that can make their own food. In addition they are more or less distinctive in their rigid cellulose cell walls and, in general, their lack of locomotion. Within Plantae we have the thallus plants, i.e., without roots, stems, or leaves; vascular plants containing fluid carrying ducts; ferns; and seed bearing plants including conifers and flowering plants, etc.

We are most concerned with Animalia, organisms that cannot make their own food, are largely mobile and capable of quick movements, and can fabricate collagen, the most abundant protein in animals. In the division in which Protozoan-Protista are included within Animalia, the multicelled animals are placed in the taxon called Metazoa.

The kingdoms are divided further into Phyla, which distinguish radically different body plans or organization. There are many animal Phyla, which include sponges, jellyfish, flatworms, roundworms, segmented worms, starfish, mollusks (snail, octopus, clam), arthropods (centipedes, insects, spiders), and chordates, already mentioned. Subdivisions of each phylum are illustrated by the listing for mankind.[16] As most chordates are vertebrates we have omitted this subphylum.

| | |
|---|---|
| Phylum | Chordata |
| Class | Mammalia |
| Order | Primates |
| Family | Hominidae |
| Genus | Homo |
| Species | Homo sapiens |

## Mammals

From the listing above we see that the next category in our developmental history is the mammalian class. Changes to protect embryonic development were described earlier by the appearance, first, of the amphibian strategy of a two part — water/land-life cycle, and then the reptilian, cleidoic-amniote (shelled and membrane enclosing) egg. These measures were carried an important step further with the appearance of mammals, about $1.50(10)^8$ BP. While the distinguishing feature of mammals is that they maintain their young by feeding from mammary glands developed for that purpose, other changes introduced by mammals also make them a most successful class.

Almost all common animals with which we are familiar are mammals with the further subclassification of Eutheria. In the Eutheria, fetal development is within its mother's body, with the embryo in its fluid filled amniotic sack being sustained by a placenta. Placental nourishing allows a prolonged period of fetal development. The young is live-born and, while it is generally not without capability, nursing from its mother allows continued, enriched, growth and development and offers a time of parental protection and education to improve survival. For example, among modern mammals it is known that many carnivores, such as the lion and the wild dog, teach their young to hunt, and that grazing animals teach their young to stay in the herd and forage cautiously. The ability to learn behavior is promoted by a long external nurturing period during which the mammal brain reaches a size that is proportionally larger than the brains of other forms of life. This is particularly pronounced in humans where more than a dozen years are needed to reach puberty.

Sidebar discussion: Beside Eutheria, there are two other subclasses of mammals. These are the egg laying Monotremes, now represented entirely by the duckbilled platypus and spiny-anteater of Australia and New Guinea, and the Metatheria, or marsupials, which are presently almost entirely resident in Australia as well. The Metatheria are mostly non-placental animals whose young are delivered, still in an embryonic state, to make their way (generally unassisted) to a maternal pouch where they attach themselves to mammary glands. There they remain until quite mature.

The geographic separation of these mammalian classes resulted from the breakup of the great southern continent, Gondwanaland.

From the time that the earth's crust first formed it has been in motion, driven by flowing matter in the mantle. It is believed that deep mantle matter is heated by the underlying core, creating convective cells in which columns of magma rise, move along the surface, carrying the "rigid" crust with it, and then descend. For centuries cartographers and astute observers have recognized that continental outlines fit together fairly well, like a broken puzzle. However, belief in the immutability of global structure created resistance to the supposition that they had, indeed been a joined entity. The theory of Continental Drift, now referred to as Plate Tectonics, was given its first modern, reasoned, and documented basis by Alfred Wevener in 1915. Supportive evidence has been found from the continuity of rock formations and mineral deposits across what are now widely separated continental borders, and from the continuity of glacial markings and fossil lifeforms.[17]

Approximately $5(10)^8$ BP the land masses of North and South America, Europe, Asia, Africa, Australia, and Antarctica formed a super- continent, called Pangea (meaning the entire earth), located in the southern hemisphere. By $2(10)^8$ BP it had separated into two daughter continents. Gondwanaland, still in the South, consisted of the present South America, Africa, Antarctica, Australia, India and the island of Madagascar. Laurasia, now in the North, consisted of North America, Greenland, Europe, and Asia. Since that time the continents have continued to drift to their present positions. This is illustrated by the separation of South America and Africa. From about $2(10)^7$ BP magma has been rising along the mid-Atlantic rift, pushing apart their continental plates at approximately an inch per year.

At the time of the initial South continental separation, Australia and South America were largely populated by marsupials rather than placentals. It was not until much later, between 3 and $4(10)^6$ BP when South America and North America became linked by a regional elevation that produced the Central American bridge, that placentals became dominant in South America. However, the Australian isolation continued until very recent, historical, times when humans introduced placental animals there. The separate evolution of marsupials illustrates an evolutionary principle called "convergence" according to which common features are found in animals with different histories of descent, when those features offer survival and efficiency advantages under the given ecological conditions. Thus, despite their isolation one finds marsupial animals that closely resemble the wolf (now probably extinct), squirrel, mouse, anteater, mole and, in South America, a marsupial tiger (now extinct).

Other examples of convergence are well known. Among them is the similar swimming form of fish and of whales, since the latter derive from land mammals that have returned to the sea. Similar eye structures are found in a host of independently evolved life forms. In particular, the eye of the octopus, a mollusk, is almost identical to the mammalian, human eye.

The reproductive method of mammals is a costly gamble taken by its participants. Not only does it involve parental risk in warding off predators during the vulnerable periods of pregnancy and infancy, but it involves a considerable maternal expenditure of energy during gestation and nursing. Analysis has indicated[18] that the size of mammalian offspring is limited by the adult female metabolic rate and ability to supply the needs of fetal and infant growth. This investment of energy, time, and self has been optimized by a switch of reproductive strategy. Instead of producing many offspring, some fraction of which survive on their own, mammals have few, who are carefully assisted and taught how to survive.

Mammalian evolution also brought other important anatomical and physiological changes. Foremost among these is the fact that mammals, like birds, which also evolved from the earlier reptiles, are endothermic, or warm-blooded. By this is meant that they regulate their internal temperatures over a large range of external conditions. (It has been suggested that some dinosaurs may have had some similar adaptation, but the evidence for this is meager). This makes them more efficient in supplying the brain and muscles with oxygen and is a tremendous advantage in the struggle for survival under conditions of seasonal and millennial changes. It has allowed these animals to expand their geographical ranges from the tropics to the arctic. However, these advantages are not gained without some sacrifice. While the elevated mammalian temperature leads to greater vital activity, a good portion of that activity is expended in meeting the physiological demands for more energy, i.e., gathering food.

In addition, mammalian legs are set under the body and fully suited for rapid, efficient ground travel. Recall that amphibian legs evolved from fins used for walking, with the result that they sprawl sideways, and the reptiles present a mixed anatomy in this regard. Also, mammals have elaborated teeth that are suitable to chewing a great variety of foods, e.g., hard, soft, gritty, fibrous, etc. They are adaptable to a great variety of living styles.

The development we are following continues through a particular Order of mammals, the Primates, that first appears in the fossil record about $7(10)^7$ BP. Judging from that record and from the presently living primates, these animals have a particular plasticity that suits them well to adapt to changes of environment;[19] overall they are physically unspecialized. Their thumbs and usually the great toes are opposable, i.e., can be bent toward the other fingers and toes, for climbing and grasping tree limbs in their natural arboreal homes. The Anthropoid apes often assume an erect or semi-erect posture. It may be noted that an enhanced olfactory ability is probably less useful in the treetop setting than on the ground and primates are generally less keen in this regard than are land-dwelling animals. Most have binocular vision and color vision for estimating distances and distinguishing objects in the random treetop environment. It is conjectured that the complexity of this environment, and the stimuli it presented, demanded rapid, adaptable, interpretive skill that promoted the initial brain and mental development that became characteristic of their descendants.

## Mankind

Our study of "man" begins with the Old World higher primates. The populations of Africa and South America have been separated since Gondwanaland, was split. We have already mentioned the effect this had on evolution in Australia. In addition, during that elapsed time changes occurred in the populations of Africa and South America. For example, South American primates bear the classification Platyrhini, referring to their flat nose and laterally directed nostrils, while those in Africa are Catarhini, with curved noses and downward directed nostrils. This difference, minor in itself, is noted here not only because mankind obviously belongs to the latter order but because of what it tells us about developmental evolution.

We mentioned earlier the basic evolutionary principle that population changes occur in directions that lend greater opportunities for survival and reproduction. In this case, however, are two populations, an ocean apart, that have developed different facial structures under seemingly comparable conditions and without any obvious advantage for the difference. While the evolutionary paradigm is doubtlessly true, this example illustrates that the reason for a particular change is not always clear. The altering of one feature may be an incidental accompaniment of some other more valuable survival or reproductive shift, since genetic mutations can and do cause more than a single character variation. Other examples are the loose linkage of eye color with skin color in humans and the development of flight in birds whose feathers probably first served merely for warmth. Or evolutionary change may result from a random, dominant genetic mutation that is otherwise survival-neutral. Campbell[6] points out that:

> Each of the twenty or so species of African antelopes has a recognizably different horn ... the differences are not adaptively significant. Instead the various forms seem simply to be alternatives.

While it is appropriate to examine evolutionary shifts in terms of the survival-reproduction rule, we must remain open to the possibility of unexplained, at least to us, or opportunistic changes.

Before continuing, to clarify some of the distinctions to be made later, let us return to the classification and development given earlier. This is done in somewhat more detail[12] on Table 3-1. Primates are divided into two suborders: Prosimii (mostly consisting of small, tree dwelling animals such as Tarsiers, Loris, Lemurs, Tree-shrews, etc., which we will not consider) and Anthropoidea or higher primates. Anthropoids have the two infra-orders, Platyrhini, and Catarhini, just discussed. This constitutes a minor distinction and is omitted from the table. There are two Superfamilies within the

**TABLE 3-1**
TAXONOMY AND PHYLOGONY OF MAN

| | LINE OF DESCENT | OTHER GROUP MEMBERS |
|---|---|---|
| Class: | Mammal | |
| Order: | Primate | several |
| Suborder: | Anthropoid | Prosimii |
| Superfamily: | Hominoid | Cercopithecoid |
| Family: | Hominid | Pongid Hylobatid |
| Genus: | Homo | extinct |
| Species: | H. sapiens | extinct |

Catarhini: Cercopithecoidea, the Old World monkeys or Lesser Apes, and Hominoidea, consisting of apes and man. In turn the Hominoids have three families: Hylobatidae (whose living representatives are gibbons), Pongidae, or Great Apes (present members are orangutans, chimpanzees, and gorillas), and Hominidae (the family of man). The stages of separation in this last phylogony are not unequivocal.[2] On the basis of recent DNA studies, described in Chapter 1, some now consider the chimpanzee and gorilla to also belong in the Hominid family, and to have separated somewhat later. In any event it is the genus *Homo* in which we are ultimately interested.

The lower primates are tree dwellers. According to conventional wisdom, the prehistoric descent from the trees, by our ancestors, was probably initiated by new weather patterns. During the period of separation of the South American and African continental plates, the Indian plate had rapidly moved northward to collide with Asia, thereby pushing up the Himalayan mountains.[20] In addition, from $5\text{–}6(10)^6$ BP there was a great uplift of eastern Africa, along the great African fault line. About this time also was the Messinian crisis, the sealing off and drying of the Mediterranean basin. As a result of these changes there was major shift of air movement and of weather patterns throughout the world. In south and central Africa the tropical forests retreated and lush vegetation became more scarce. Apes found themselves on the ground with greater frequency, to find fruit and low shrub and root foods and, for those living near forest margins, to travel between the now separated wooded areas. Over the long term, this led to bipedal locomotion.

Climate may not have been the only source of the change to walking on two limbs instead of four. It has been suggested that a major cause of infant mortality among tree-living primates is the loss of infants who fail to cling to their mothers and fall to the ground. A subspecies in which the females begin to hold their infants would leave more offspring to the next generation but, as a result, would be less agile in the high branches and would likely pass more time on the ground. Holding an infant with one fore-limb, and retaining one for picking roots and berries requires bipedal locomotion.

While some stages of the resulting physical changes are obscure, comparison of humans and apes and study of the available fossils, indicate the host of modifications that took place.[21,22] Ground life in the open forest regions is fraught with danger from stronger, faster carnivores. A more erect posture for early warning, and a flat-soled foot, for more rapid escape and walking on hard surfaces, are survival adaptations in these circumstances. A major change in this regard is the forward alignment and relative size change of the large toe that, in man, takes the full body weight and acts to push off each step. With this, and arcing of the foot and spine, a flexible springy stride is possible. Longer and straighter leg bones and modified muscle strengths produce the capability of relative speed with a stride that also allows tireless, long distance travel.

Development of an upright, S-shaped spinal column, rather than it having a single forward slope, and backward rotation of the hips to allow the longer stride and to support the body weight. had an important secondary effect. To balance the head directly above the spine and pelvis, the cranial opening of the spinal cord, the foramen magnum, moved beneath the skull rather than entering it from the rear, as in the apes. As a result, strong supporting neck and shoulder muscles were no longer necessary and the heavy, bony ridges for their attachment at the rear of the skull, disappeared.

Upright posture and walking freed the arms and hands from their use for support and travel, and they developed in ways suitable to other functions. In humans the arms and finger bones are straightened and, in particular, the thumbs are longer so they can be touched to the other finger tips, allowing for fine manipulation as well as a firm grasping.

Changes also occurred in jaw structure. Possibly these resulted from different eating habits or from the substitution of the hands for holding, food gathering, and protection — in place of powerful jaw muscles and prominent teeth. The front teeth became smaller and vertical, and the rear, grinding teeth decreased in size. The entire skeletal structure became less massive and, in particular, the jaw has less bulk. Recession and lateral rounding of the jaw changed its frontal structure and a reinforcing ledge of bone, the simian shelf, has disappeared. This shelf is the point of attachment of the tongue in Apes, so a new, ultimately more flexible configuration has developed for the tongue muscle. Also, in some species the jaw muscles are attached to a mid-cranial sagital crest and to prominent, laterally spread cheek bones. These features decreased in prominence, with the expanding brain case providing a sufficient area for attachment of the now reduced musculature. Similarly, the eye ridges, which also acted to distribute the great stress resulting from hard biting and chewing, are reduced. It is difficult to tell which is cause and which effect but these developments reduced the bulk of bone mass that surrounds the skull. They are accompanied by an upward and lateral expansion of the brain case to hold a larger brain, the final and most important distinction of humanity.

Of course all these changes did not occur simultaneously but eventually the departure from other Hominoidea was great enough that a different Family name, Hominidae, is necessary to describe the new creatures. Several different fossil variations, corresponding to different genera of Hominidae, have been found.[21] Dating about $10(10)^6$ BP are remains of Ramepithecus, possibly the earliest Hominid. The Australopithecenes, so named because the stem austral (southerly) denotes their first discovered fossils in southern Africa, appear between 3 and $5(10)^6$ BP. These were small (about 4 feet tall) and probably ape-like in facial appearance and structure, with mostly upright posture.

From about $2(10)^6$ BP one finds fossil remains in Africa of the genus "Homo" represented by *Homo habilis* or handy-man. This was also a small, erect creature with a low vaulted cranium and, likely, a broad flat face. But *Homo habilis* purposely shaped stone, and probably shaped animal bone, to make a sharp cutting edge. By about 1.6 million BP *Homo erectus* or upright-man, has appeared, with the ability to make such tools as a hand-axe and, from some fossils found associated with burned animal bones, to use fire. *Homo erectus* had a smaller cranial capacity and an intermediate facial development relative to modern man but was a most successful creature; for over 600,000 years, races of *Homo erectus* spread from Africa throughout Asia and Europe.

The Australopithecenes and these early species of *Homo* were progressively closer to modern man. However, while the matter is still debated, it is generally held that they constituted separate lines of descent of the family of Hominids, i.e., they were not our direct progenitors.

## Homo sapiens

The physical strength of present Hominids, i.e., of mankind, is generally not equal to that of their nearest living cousins, the Pongids or great apes. However, humans regard themselves as superior by virtue of their minds, arising from a larger, more complex brain. The Hominoids have larger brains among the mammals, and the human brain is largest among the Hominoids.[23] In fact, several whales and elephants have bigger brains by factors of four to six. A better comparison is the ratio of brain to body size, in terms of volume or weight, as a measure of "encephalization". Even this may not be useful if the species being compared are too diverse, and even for narrow groups, the data are often ill-defined.

For some time it was believed that brain weight of mammalian species is proportional to body weight to the 2/3 power. This would imply a relation between surface area (proportional to the square of length) and body weight (proportional to the cube of body length). Any such overall relation must be interpreted with caution because all parts of the brain may not change uniformly. In particular, the human cerebellum, which is generally associated with higher abilities, is a thin shell with a highly convoluted surface. However, more recently it has been found that the correct power for this relation of brain to body weights is closer to 3/4. This implies an overall energy limitation to brain growth since metabolic rates increase with this power.[24]

While no correlation between brain size and intelligence has been found within a species, the difference between species is significant. It has been pointed out[13] that development beyond *Homo erectus* presented a problem. The size of the birth canal and, as indicated above, the energy potential of the primate mother limit the fetal birth head and brain size to about 350 $cm^3$. In most primates, brain growth continues after birth, typically doubling in weight. As the brain of the subfamily Homo has increased beyond that of *Homo erectus*, a way must have been found through this impasse. Anthropologist Steven Jay Gould has proposed that the solution evolved by nature was to take advantage of preadaptation, a process we have already mentioned; here it is the extension of a tendency already discernible in the Pongids.[25] This is accelerated fetal growth and decelerated fetal development so that the body and brain of human infants are larger (twice as big) but less mature than expected from the gestation period. Human infants are born with essentially fetal brains and between birth and adulthood, along with body maturation, the brain expands further, by an additional factor of four.

Physically and mentally human infants are more helpless and demand considerably more parental care than is the case with other primates. From comparisons of the fetal and infant growth and maturing of humans and apes, it has been suggested that human gestation should really be considered to last 21 months from conception. The last 12 months are spent extra-utero, subjecting the infant to intensive stimulation and learning opportunities as it matures to a "proper" primate birth state.

This process of delayed development, known as Neoteny[21,25] — a premature termination of gestation — is consistent with a number of other unique human features, as well. In the early and mid-nineteenth century it was observed by K. E. von Baer and others, that the embryos of higher animals pass through a succession of stages resembling the corresponding embryonic stages of lower animals. An example of this is the pharyngeal pouches of the human embryo, i.e., embryonic gills, which are then converted to form the inner ear, thymus and parathyroid glands. Similarly, the jaws of primitive vertebrates consist of many bones and while these are displayed by the developing human fetus they mostly fuse into the single primate jaw, while some translate into the ossicles of the inner ear. In view of this fetal development and of the "premature" termination of human gestation, it has been suggested that many human features can be taken to be prematurely interrupted pongid fetal growth characteristics. For example, during most primate fetal development the spinal foramen, the opening of the skull through which passes the spinal cord, is near the center of the base of the cranium, its position in humans. In the primates it shifts to the rear only late in that development. The appearance of bony eye ridges and forward projection of the face also occur late in the growth of fetal and young simian primates, as does the development of body hair. Similarly, the position of the great toe is more parallel to the others in fetal apes and moves laterally near the end of gestation.

By whatever mechanism and through whatever causes, with these changes we have the species *Homo sapiens* (wise man). Archaic (i.e., non-modern) forms of *Homo sapiens* first appear in the fossil record[17,26] between about $5(10)^5$ BP and 1 to $2(10)^5$ BP. Among these was *H. sapiens neanderthalensis*, Neanderthal man, who had a large brain in a low and elongated skull, rather than the vaulted skull and vertical forehead of modern man. These changes did not occur uniformly in all places at the same times, and it is noteworthy that with the appearance of each succeeding type of Hominid, the previous one seems to disappear abruptly. It has been suggested that this resulted from absorption of one into the other through interbreeding but most modern scholars believe that extinction resulted from food and living competition, or from war and murder, possibly motivated by that competition.

## Homo sapiens sapiens

One additional feature was needed for sapient man to achieve the potential promised by his expanded brain. To be truly thinking, it is necessary to have the tools and symbols with which to formulate thought. Ideas do not exist in a vacuum; their organization in the mind demands categorizing the objects, actions, and concepts being considered. Particularly if these are not simple, they must have symbolic representations that bear the import of their varied features and behaviors, and they require means of interrelating and associating diverse aspects. These functions are served by names, words, and language structure. Therefore, before anatomically modern *Homo sapiens* can be considered to be truly thinking man, i,e., to be "*Homo sapiens sapiens*", he must have developed speech.

Now the ability to make sound does not constitute speech. The vocal signaling that transpires between the members of lower animals is communication of a sort, but is generally not adequate for other than the simplest message such as warnings of danger, calls for mating, etc.[27,28] Undoubtedly archaic *H. sapiens* had refined such signaling and communication but it is a much argued question[29] whether they were capable of any form of speech as we know it. Dr. Philip Lieberman[30] and other linguists, physiologists and anthropologist have compared the skeletal forms of humans and existing primates with fossil remains.[31] They have been particularly concerned with the relative structures and positions of the mandible, the thorax, the foramen magnum, and fine structures in the base of the cranium and the palate. From details of the fossil bony parts they have been able to infer knowledge about the adjacent soft tissues. Such tissues are not left in the fossilization process but in life they accommodate to, and leave impressions on the hard parts. These scientists have concluded that it is highly unlikely that even archaic *Homo sapiens* had the mouth and throat structure for speech.

In modern man the larynx has moved lower in the throat, the tongue has thickened and become capable of refined control of the pharyngeal air canal above the larynx, and the velum, extending behind the hard palate, has come under refined control. Only man moves his tongue when making sounds. These adaptations to speech were not made without sacrifice. Improved precision and rate of vocal transmission are accompanied by several features that can be life threatening. To mention two of these, Lieberman notes that: 1) Narrowing of the air canal makes for less efficient breathing, e.g., as swallowed food must pass over the opening to the laryngeal air canal, posing a greater likelihood of choking. Unlike our pongid relatives and newborn human infants, we cannot isolate the larynx and esophagus to swallow and breathe simultaneously. 2) Rounding and reduction of the mandible has crowded our teeth, increasing the likelihood of compaction and infection.

The evolutionary shift toward the development of speech required a concurrent neural development for comprehension and articulation of language, as well as for the refined, rapid, coordinated muscle control that speech involves. Indeed, a mapping of the centers of functional control in the brain does disclose appropriate areas of specialization, especially in its left cortex.[32] Immediately adjacent to the left motor control region, particularly to that for the face, tongue, jaw and throat, are Broca's and Wernicke's areas. These appear to moderate the production and the comprehension of language, respectively; they are connected, and coordinated, by a bundle of nerve fibers called the Arcuate Fasciculus. Wernicke's area, in turn, is adjacent to the primary auditory area and is linked to the primary visual area through the angular gyrus, which mediates between visual and auditory signals.

While some cerebral asymmetries are found in many species, and associated with various abilities, they are hardly comparable to those for humans.[33] The specialized areas described just above have no clear counterpart even in lower primates; their vocalizations arise from the more primitive, more deeply situated, limbic regions of the brain. Nor do these regions have counterparts in the right hemisphere of the brain. Surrounding the right side motor areas are centers that appear to specialize for geometric visualization and music. It is the asymmetric, bilateral specialization, which is most pronounced in right handed people, that forms the basis for an explanation of the origin of speech.

The question raised by such associations is to explain how they came to be, i.e., their history and especially their evolutionary value. The ability to communicate by speech lends considerable primitive survival advantage, say when running down animal prey, or when defending against warring tribes, or when teaching survival techniques. However, it is hard to visualize conditions that are conducive to development of the necessary, disparate skills in early *H. sapiens.* This is unlike many other complex adaptations for which the early and intermediate states themselves offer distinct advantage. For example, the eye, with its associated nervous and brain systems, is a sophisticated instrument. It is capable of focusing and representing true images, adaptable to an extremely wide range of intensities, most sensitive to motion and changing edges and shadows, etc. In this case it can be understood that simple photosensitive tissue, offering the ability to distinguish bright from dim, and shadow from not, has survival value over the absence of such tissue. Once present, this tissue can readily evolve, with its interpretive nervous backup, to allow more refined and wider range properties. Evidence for the universal advantage of vision is provided by its wide dispersal among diverse life types, albeit in modified forms.

William Calvin[34] has proposed a scenario to explain the evolution of speech as an accessory development to another ability that had immediate survival value for the primitive hunter-warrior. This is the ability to throw stones accurately. According to Dr. Calvin, our ancestors, like present apes, had no distinct group handedness; left and right were equally likely in any individual. However, when primordial mother descended form the trees to scavenge meat or hunt berries she became more susceptible to the actions of predators. Over time she learned, as we know today, that the infant she carried was more likely to remain calm when held on her left side — closer to, and better able to hear and sense her beating heart. Her right hand was then free to do the more precise work of digging, picking and, when threatened, throwing. Now one-handed throwing is an entirely different exercise from the two-handed heave used by most apes. As we know, it is capable of greater speed, range, and accuracy.

However, these attributes are not easily attained. Extremely refined timing and anticipatory control (on the millisecond level) are required of the sequential action of shoulder, arm, hand and finger muscles. Another requirement is for equally refined hand-eye coordination, if the missile is to be accurately launched so that a relatively distant, moving target (such as a small animal being hunted) is hit with enough force to stop it. Those individuals whose genetic make-up made them, however slightly, better able to do so were more likely to thrive and pass that ability to their offspring. Over the millennia that ability was enhanced and the motor area of the left brain, which controls fine sequential motion of the right side, grew and became more significant. It is probably no accident that the site of this area is also adjacent to the motor control area for the face, tongue, jaw and throat. Calvin suggests that improved ability in these latter areas led to the fine control of expelled breath and from there to speech. According to this suggestion, our ability to speak may have developed as a side effect of the more mundane, but at the time more important, ability to attack at a distance.

This lead to right handedness in most of the population, and enlargement of this brain site may well have been the driving force for general brain growth. Other areas, on both sides, developed different special abilities for interacting with and interpreting the world, eventually leading to the brain of Homo sapiens sapiens.

It is interesting to note one other piece of evidence for the specialization that relates the unique appearance of speech ability and the expansion of the cerebrum, with the host of other capabilities of *H. sapiens sapiens.* Very recently X-ray tomography has been applied to fossil skull remnants of *H. neanderthalensis*. This involves a penetrating computer analysis of extensive X-ray data from the sample, that discloses details of internal structure. It has been found that the ossicles of the inner ear, the bones that conduct sound from the ear drum to the internal, sensory nerves, have a different construction. It follows from this that the perception of sound and, conversely, the ability to make sound, will be different, as will be the necessary interconnections in the brain.

The human brain is vastly more complex than can be explained from this scenario alone. It is not merely a control mechanism for refined mechanical action and speech control. The human mind is adaptable and creative, it displays consciousness and intelligence, and it fosters emotions and culture. There is no natural competition that could have promoted and tested these developing mental powers sufficiently to cause the high level of sophistication they now display. Lower animals exhibit awareness, intelligence, purposeful action, emotion, communication, and learning, but these are on an entirely different level than in humans. It has been suggested that the only evolutionary process that could have promoted such vast and sharp growth of the mind is intraspecies conflict. Competing with the closely matched, rapidly changeable abilities of fellow *H. s. sapiens* for control, advantage, sex, etc., has provided a form of mental, natural selection. It has also sharply honed the abilities, some of which are also seen to a lesser degree in animals, of deception, seduction, lying, etc., along with mental visualization, inductive reasoning, planning, speech, etc.

> In fact, anthropologists have formally proposed that man's emergence was precipitated by the development of a breeding system that transferred an overwhelming influence over survival to the hands of members of his own species. Deliberate parental breeding and social behavior replaced the external environment as the main determinants of fitness of phenotypic traits. Conscious behavior became the driver for primitive man's evolution.[35]

In any event, analysis of fossil evidence from the mid-east indicates that between 90,000 and 100,000 BP modern mankind appeared on the scene.[19,22,36] The location of these fossils is consistent with the "out of Africa" scenario mentioned in Chapter 1 and in the next chapter. It is also interesting to note that these early migrations appear to have primarily turned east after crossing the Levant, leading to early Asiatic evidence, such as Java man. It was not until about 40,000 BP that "Cromagnon" man appeared in Europe.

# 4: POST-BIOLOGICAL EVOLUTION

## Civilization

The record of the Earth shows many extended periods of extreme climatic conditions, with a long history of large-scale glaciations. Direct evidence of the successive glacial periods is given by characteristic geological rock markings and deposits left by moving and melting ice. Additional evidence is supplied by fossil remains indicating the types and living domains of animals and plants that inhabited the glacial regions. Even subglacial regions experienced climatic changes and the successions of alternating rainy and dry periods affected the flora and fauna and are therefore recorded in fossil records.

The most recent geologic period, the Pleistocene, is particularly well known to have had a succession of ice ages.[1] Its first glaciation began 600,000 BP, a second about 500,000 BP, the third about 250,000 BP, and the fourth about 120,000 BP. The last of these had fluctuations of severity with the times of greatest glacial extent about 110,000, 72,000, and 24,000 BP. The period of this fourth Pleistocene ice age coincided with the appearance of *H. s. sapiens*, whose migrations out of and away from Africa occurred, primarily, during the interglacial periods.

In these migrations, as in the previous and subsequent activities and changes of life style, early man doubtlessly moved in groups. Like our primate cousins, we and our ancestor are social animals, living and functioning in social and survival communality with our fellows.[2] In such animal groupings, one generally finds a hierarchical order. Lower ranking individuals defer to those of higher rank in nearly all activities, such as food distribution, social interactions like grooming, selecting resting locations, sexual access, etc. Status is generally inherited and individuals who try to exceed their position are frequently punished by beating, expulsion from the group, or even by death. Referring to the social enforcement of order among fowls, this class positioning is often called a "pecking order." Investigators who study these groups point out that maintaining the social order reduces tension and inefficiency. Artificially disrupting it, for example by introducing new members or altering individual markings to obfuscate the recognized order, leads to chaos and deficient maintenance of the entire group.

In keeping with the evolutionary imperative, that long-term survival means promulgation of ones genotype to future generations, we would expect it to be the natural tendency of individuals to promote the survival of only themselves and their own progeny. However, in many animal groups the *common* interest is maintained by all its members, often without regard for this consideration. In some the very young receive care and are nursed by any available female, and group defense is often at considerable personal risk.

This behavior begged for explanation until William D. Hamilton clarified a mechanism of group coherence by proposing the principle of "kin-selection" and the ability of "family recognition."[3] These are displayed by nearly all social creatures. This principle postulates that genetic survival is the basis of collective group behavior. Since primitive groups are extended families, the genetic composition of kin is close enough to ones own that there is hereditary advantage to family as well as personal gene survival. It leads to a form of selfless behavior, altruism if you will, in which an individual will sacrifice some of its own self-interest to promote the group well-being. Of course these are not reasoned responses. Rather, they are automatic reactions, conditioned over millennia by the necessity of group fitness for individual survival, and by the individual's receptiveness to social pressures so he is accepted by the group.

With the cognitive ability of mankind, group affiliation had been extended from the family to the tribe and from the tribe to the larger group, be it religious, ethnic, geographic, or national. In any event, intragroup cohesiveness and social interactions and intergroup competitions provide the background elements for the post-biological evolution we wish to describe.

The Quaternary period witnessed humanity's transition from intelligent animal to participant in a civilized society. In Chapter 1, we divided the Pleistocene Epoch into lower, middle, and upper ages, as determined by glacial intrusions. These geological divisions are paralleled by stages of evolution of the genus *Homo* and in the present discussion we will use alternative notations for that division, in terms of the process of civilization building.

- The Paleolithic or Old Stone Age might be considered to date from 1.8 to $2(10)^6$ BP. The earliest technology appeared then, primarily the development of *H. habilis.* It is known as the Oldowan and consisted of breaking stones and selecting those which were sharp and useful. Starting about $1.5(10)^6$ BP, and associated with *H. erectus*, the characteristic tools are labeled Acheulian after the French site where they were first discovered. Similar tools were subsequently found throughout the intercontinental ranges of this species and frequently bear other reference names. These stones were chipped on both sides to make a sharp edge and rounded on the opposite edge, as for gripping. They were likely used for cutting and scraping.

- As the earth warmed following the glacial period, the formerly frozen water from the glaciers became available in the form of rain and in rivers. Plant and wildlife proliferated and the new tools could be used to more successfully hunt, forage, and scavenge. The Mesolithic, or Middle Stone Age, saw the development of more refined and sophisticated tools. The Mousterian culture, about 35,00 to 40,000 BP and primarily associated with *H. neanderthalensis*, produced flint for scrapers, projectile heads and knives.

It is uncertain whether the Aurignacian tradition, which followed shortly, represents an evolution from the Mousterian, or the influx of more modern species of *H. sapiens*. This age saw the development of the spear thrower, which was likely mankind's earliest mechanical device, giving him an effectively longer throwing arc and additional flexor joint with which to impart greater thrust to his missile. About 15,000 BP the bow and arrow was invented and its use spread rapidly. Domestication of the dog also dates from about that time, possibly initiated by hunters bringing home orphan or stray pups to be family pets. Other animals, likely those that grazed from the gleanings of harvested fields, came under human protection and then herding. Rafts and dugouts were developed for crossing water, but they were not fitted with sails until some time later.

The abilities to conceptualize, to appreciate the passage of time and events, and to contemplate one's own existence and death, lead to attempts to understand the apparent uniqueness of mankind within the world in which it was embedded. The great scale of nature and its awesome forces and mysteries lead to spirituality and worship and, very early, to preparation and burial of the dead. Expressions of magic are manifest in rock drawings and in the ancient cave drawings found in Spain and southern France. These sentiments gave rise to shamans and witch-doctors who claimed special skills in guiding prayers to the greater powers, in curing the sick, in foretelling the future, etc.

- The Neolithic, or New Stone Age is marked by the first processes that can be said to have initiated modern civilization. It began between 11,000 BP and 9,000 BP in what is called the Fertile Crescent of the Middle East;[4] the Levant is a nexus of population flow between Africa, Asia and Europe. Approximately 16,000 BP the last ice sheet began to melt, causing a shift of rainfall and temperate climate from the North African to Middle Eastern regions. As a result , this latter developed fertile river valleys and well-drained hillsides; it was a natural site for the initiation of agriculture. At first those fruits, roots, and cereals were tended that had formerly been gathered from the wild. Gradually, with the understanding of seed propagation, improved crop types were selected.

It is noteworthy that the early selection of seed grains by human farmers induced a major change in these food crops, as though they were molding themselves to the farmers needs. The cereals: sorghum, rice, maize, wheat, millet, barley, oats, and rye are related members of the family of grasses. In their wild form, they generally have small seeds that mature early and scatter themselves and penetrate the soil. These features make cultivation difficult . It has recently been shown[5] that a small number of genetic loci control these behaviors and by selection of a few appropriate mutants vastly improved varieties were developed. These have longer maturities, and thereby fuller grains, and can be harvested before the seeds scattered, leading to more efficient harvesting. It is remarkable that these parallel changes occurred in the different grasses independently in widely separated parts of the earth, each over a period of perhaps only a hundred years.

Specific cultivation came into being using digging sticks and bone or flint-tipped hoes. These developments produced steadier food supplies as well as accessory materials such as fiber from flax, and wheat and barley shafts for thatching houses. Farming can sustain a much larger population density, and the necessity to care for and tend to cultivated lands led to more permanent housing in horticultural villages. Animal domestication continued in these villages, probably as a secondary activity, with the goat, sheep, ox and pig as suppliers of meat, bone, and skins. It was not until much later that some of these were utilized as beasts of burden.

Independently of farming, and perhaps only slightly later, pastoralism arose. Whereas farmers moved occasionally, as their fields became depleted, shepherds and goatherds had constantly to seek new pastures. These nomadic tribes served as communication links and traders between diverse areas. By 8000 BP they were spreading the use of pottery. With the initiation of the bronze age about 5500 BP, they served as vehicles for the spread of metals with their considerable advantage in making tools and weapons.

We cannot dwell on these two distinct lifestyles, agricultural and pastoral, which developed different regard for nature and different attitudes toward man's place in it. Our interest is in emphasizing certain aspects of technological advance, and we are forced to skip important features of the development of civilization. These include different farming techniques, such as slash and burn and contour plowing, the invention of the wheeled cart, sociological and ethnic groupings that developed and the consequent warfare between them, and varied religious beliefs, though we will shortly return to the role of the priests. In this regard, we should also note that parallel development took place, apparently independently, in other parts of the world, e.g., Egypt, the Indus valley, China and, later, in Central and South America. "All agricultural societies developed during the last ten thousand years of humanities hundred thousand years on earth."[6] Our emphasis on the Levant is due to its being, according to available evidence, the earliest to witness these changes.

Both farming and pastoralism were unreliable, survival level activities, in spite of their great contributions to the welfare of mankind. Even with the technological advances made through invention, the use of metal tools, and of animals for heavy draft, both were at the mercies of disease, pestilence, flooding, and drought. The fight against plant disease and infestation was based on such techniques as migration, rotation of crops, and successful breeding and cultivation. It was not until the present century that the use of chemicals could be said to have brought this problem under control. The need for, and lack of, a stable water supply was dealt with between 6000 and 7000 BP. In what must have been a monumental, cooperative effort by craftsmen of different training, by farmers, herders, military leaders, religious men, etc., the swamps of the Tigris and Euphrates river valleys were drained and made livable. In the area, later known as Sumer, sluices were constructed to redirect surface water and Dams were built to restrain and store the waters of periodic flooding. The swamp land was filled and raised, and irrigation began. Recent archeological studies in Bolivia give a more detailed picture of how this might have been done, albeit some 4000–5000 years later.[7] The duration, scale and effort of this endeavor in Sumer left a deep impression on the memories of the people, being reflected in their mythology and, possibly in the biblical statements, in *Genesis*, of separating the waters and the dry land.

The construction and use of irrigation systems requires the joining of knowledge and effort in a technological, community effort. It implies pre-existing social, economic, agricultural, and engineering sophistication and leadership. This is, perhaps, the most prominent example of the new type of communal structures that were created by grouping individuals with different capabilities, by forming control hierarchies, and by "institutionalizing" management and methods of interaction. By these means, the method of change and evolution shifts because the codes and means determining growth, advance, and change are now stored outside of the physical body. They do not die with the individual but remain a property of the group, to be modified or carried forward.

Subsequent change is no longer mediated by the reproductive period of the biological system; these social systems can restructure themselves very rapidly. Such communities also establish norms for the protection and rearing of children. All these features together act to remove survival, propagation, and progress from the realm of "natural selection" and to place them under the control of the community itself. In this regard M. Maxwell[8] points out that in such efforts "Society adds a new layer in evolution."

The rise of towns involved a society with common interest. These include trading surpluses for essential supplies from outside, control of water distribution, distribution of farming and grazing land, and belief in the particular god or gods which controlled that land. Each field and settlement was the domain of a god, on whose behalf the fields were cultivated, and these were served by the local shamans and medicine men. Conflicts between towns led to wars and conquests so that large areas became controlled by strong kings and governors. Merging of the various local beliefs took place and led to a new class of priests.

## Writing

It was the responsibility of the priests to ensure that the godly host and ruling kings were properly provided for, protected, and that productivity was maintained. Among other functions, they were responsible to determine the opportune times for sowing, harvesting, gelding, etc. For these they had to understand the annual calendar. Not only was the relation between the monthly and annual cycles known, but solar and lunar eclipses were predicted. We should note that these abilities did not demand sophisticated astronomical knowledge. Although it requires interpretative skill to coordinate the observations, the periods of cycles of nature and of astronomical phenomena are susceptible to direct numerical comparison.[9] Savants in several centers of civilization were able to learn the relevant patterns of repetition and to coordinate them with terrestrial events such as periodic floodings, appropriate planting times, etc.

They also had to maintain equity of production, trading, loans, contributions, and the determination of individual land boundaries. These call for a vast array of capabilities including the establishment of standards for length and weight, the ability to do geometrical and numerical manipulations, and the necessity to keep records. Permanent records and quantitative considerations demand systems of notation and these were developed about 5000 BP.[10]

This is not to say that crude forms of notation were not extant earlier. Notches carved on bones, dating from 100,000 BP, may have represented tallies, and clay containers with internal balls and external marks originate early in the period we are considering.[11] The first "writing" was in the form of pictographs — suggestive representations — made by marking clay tablets or tree bark; these are the cuneiform writings of Mesopotamia and the hieroglyphics of Egypt. In time they evolved into simpler symbols, often called ideograms, which may have barely resembled the original but that represented the object or the concept for which it stood. Such symbols were numerous, awkward, and difficult to master. Because of this, and with the needs of trading with different peoples and, particularly with the demands of varied conquerors, the images had to be converted to a phonetic form that could be read and interpreted more easily. For this the existing symbol might be taken, for example, to indicate only the first sound of its ideographic representation. Once the direct correspondence between object and symbol was eliminated, the symbol could be further abbreviated, stylized, and formalized. These steps were accomplished by the Phoenicians between 3500 and 3000 BP. The Phoenician alphabet had no vowels but it was only a few hundred years until the Greeks added this feature.

The reader should be aware that in the one, brief, preceding paragraph we have presented what scholars agree is probably the greatest intellectual achievement of mankind.[11, 12, 13] Shortness of presentation has been dictated by the desire not to become mired in what is, for us, irrelevant detail. It should not be taken to diminish the significance of the invention of writing.

For the same reasons of brevity, our discussion now skips 3 millennia, constituting most of recent history or, in view of the newly invented ability, most of "written history." During this time, as civilization advanced, its progress became increasingly dependent on writing. Legal documents and agreements had to be written for permanence and later verification. Thoughts and lessons could now pass beyond an immediate circle of listeners and messages could as well go directly to future generations and to distant parties. Knowledge and understanding could thereby be accumulated from wide ranging sources and could be compounded so that progress and innovation would proceed at a more rapid pace.

Even so, however, the pace was slow. Very few could read or write, and only those items deemed to be of greatest importance were worth being put in permanent form. For a long period these were primarily transcriptions of royal and holy works, done by hand. Information transmission was often painfully slow and of limited circulation. Mechanical printing was used but it was neither more rapid nor of better quality than hand written documents. The method used was block printing, in which relief characters were carved in blocks of wood, bound together, inked, and pressed with vellum or paper. This method was well established in Japan by the year 770, and probably long before in China. However, Asiatic pictograms were not suitable for the development of movable type, even though it was tried in China in the mid-eleventh century, and in Korea in the early 15th century. Because of societal preferences and technical problems they did not lead to continued development. These efforts were completely unknown in Europe, to which the Far East was a distant enigma, shrouded in myth and mystery.

Europe of the fifteenth century presented an entirely different tableau. Nationalism had not yet come to the fore and the economic and political scenes were chaotic. The Christian Church was in a state of unprecedented decadence and, having just ejected the Moguls, the continent was under attack by the Turks. However, the foundations of modern literature had been set by writers in the vernacular and the great Renaissance had been simmering since European exposure to Eastern culture during the Crusades. These created a demand for books, on a greater scale than ever before. To complement the need was the availability of necessary tools. Paper production had begun to replace the more costly and less available use of vellum. Oil based inks were coming into use by artists in place of water dyes which would not transfer cleanly from metal dye to paper. Metal casting was well developed for making jewelry and coins and, of course, languages were based on alphabets with individual letters.

The genius who put these elements together was Johannes Gensfleich, who used the family name of his patrician mother, Gutenberg. The hard evidence connecting Gutenberg with the invention of printing is sparse, though there is little doubt about his role.[13] Between 1439 and 1457 there are several court records of his problems resulting from borrowing money to develop a process related to printing and "the works of the books." Gutenberg invented the type mold, a method of holding uniform, metal-cast, movable type, consisting of individual letters, and he used these to press ink onto manuscript sheets. This was clear, cheap, and fast, and lent itself to making multiple printed copies. Though not his first work, his most renown is the 42 line Latin Bible, so known for the number of lines in a column page. This was published in 1455 or 1456.

> The German invention of movable type was as revolutionary for the 15th century as the computer was for the 20th. The metal letters devised by Johann Gutenberg could be assembled and printed so rapidly that books and documents were mass produced after 1450. Literacy, once reserved for the elite, spread swiftly to the masses.[14]

The significance of Gutenberg's invention can hardly be understated.

> In the cultural history of mankind there is no event even approaching in importance the invention of printing with movable types.[12]

In analyzing major levels of evolution, K. Haefner[15] describes this event by pointing out that

> ... *Homo sociales* has refined the techniques of external storage by means of painting pictures and, later, using letters and scripts. In addition to generalized neural information processing, 'externally organized knowledge' evolved which finally, by means of printing technology, became what is called "global information".

By 1457, a three-color process had been introduced and by 1500 there were several million printed books and Europe had more than 1000 print shops. Copies of religious, legal, scientific, and literary works needed no longer to be kept, interpreted, and passed along by a select few. They could be read in their original forms by any literate person. In some cases, particularly with religious works, the casting in print had the effect of fixing the content in an unalterable form, mitigating against interpretation and dissent. However, in most instances the printed word was a stimulus to new thought and scholarly study. With greater access to the thoughts and works of others, the study of nature was accelerated.

New areas of endeavor were opened and new conceptual, analytical and experimental tools were devised to probe those areas.

In Europe, almost concurrent with the appearance of printing, but not directly connected with it, was the decline in influence of religious scholasticism, mentioned in Chapter 1. The period of the Enlightenment gave impetus to studies of the new "Natural Sciences and Philosophies." After a billion years, the evolutionary product had the ability and was free to examine the processes and the forces that had produced it and could marvel and wonder over a new understanding. Many of the subsequent advances of civilization have been made possible through the discoveries and understanding arrived at through the sciences, and through their applications by science's cohort, engineering.

## The Growth of Information

The last section of Chapter 3, describing the development of speech communication, and the previous section of this chapter has emphasized the progress made through other communication means. This may be thought to have begun at least as early as the appearance of mammalian, parental survival education. However, it was certainly advanced with the ability to speak and, again, through the written and then the printed word. Our modern period has been referred as the "Age of Communication." This, of course, implies the communication of information, and forces us to recognize the tremendous, recent growth of available information and of the information processing ability. Studies have shown that this progress and growth are proceeding exponentially, i.e., with a regular, periodic doubling of numbers of published pages, of populations, of technological improvements, of stored data, etc.

It might be said that a modern communication revolution was nascent in the discovery of radio waves in the last decade of the nineteenth century.[16] Then it became pubescent with the development of electronics, particularly the invention of vacuum tube technology and its application to practicable radio transmission early in the 20th century. In its young adulthood today, auditory and visual information is communicated instantaneously between all points on the earth and almost instantaneously between earth and its spatial voyagers to the other planets of our solar system.

The digital computer lagged these early developments by only a few decades. Mechanical calculating devices have been recognized among ancient Roman artifacts, and cultures before and since have witnessed efforts to develop automatic computation systems. These range from simple adding mechanisms to complex horologic instruments showing the motions and phases of astronomical bodies. Modern, electronic, digital computing was demonstrated to be practical in 1945 by J. Presper Echert and Edward Mauchlay who constructed the Eniac computer. However, computation did not develop fully until it was joined with solid-state electronics and microelectronics.[13] These were initiated by the discovery of the transistor in 1947 by John Bardeen, Walter Brattain and William Shockley; and they have resulted in an explosion of affordable computing capability and electronic communication. Auditory, visual, and data signal transmissions are taken for granted today, and nearly every college and high-school student, and many elementary school students either own, or have access to, a personal or mainframe computer. As the power of these instruments increases the amount of information being handled and communicated is of amazing proportions.

The direction of modern society involves a close wedding of these two tools. On one hand we have the information generation, modification and handling capabilities of the computer. On the other is information communication capability using the electromagnetic spectrum.

In considering the impact of such systems as faccimile transmitters, and communication webs and networks, the historian Gertrude Himmelfarb has pointed out[17] that

> Historian are notoriously wary of the word *revolution* .... they reserve it for changes that dramatically alter the course of entire centuries .... some of us are prepared to say that we are now witnessing another revolution, ... the electronic revolution. Like all revolutions, this has ramifications far beyond its immediate context.

# 5: GROWTH AND ORGANIZATION

## Free Growth

In the previous chapters we described features of the multifaceted evolution on earth. We now wish to develop a model to use in understanding aspects of the relations between those features. For this let us establish some background by analyzing a simple case of growth. In particular, consider the common case where the change of a number or property is proportional to the number itself. If the likelihood of something happening to a member of the population (such as dying or procreating) is the same for all members then the total change (loss or gain of number) will be proportional to the total number of members that can change. Of interest to us here are cases where the change is an increase. These include the growth of capital under the compounding of interest, the growth of an embryo by cell multiplication, and the increase of a population under favorable conditions, such as tissue growth in a laboratory culture dish. Examples where the change is a decrease include radioactive decay of the population of unstable nuclei, discussed in Chapter 1, electrical discharge of stored energy, and the relaxation of active muscle tissue. Specifically, we can take the novel appearance and growth of a population with a heritable trait or cluster of traits, or else of a species in a new environment.

The individual members engage in the struggle to survive and grow or reproduce. If the traits are advantageous or the environment is nurturing, they will be successful and their prevalence and strength will increase. Let us denote the mean reproduction rate by "$r_o$", meaning that, on average, each individual produces $r_o$ offspring every generation. For example, if the average mating pair has a litter of 3 then, for the individual, $r_o = 1.5$ per generation. If the population is N at some time, then after a reproductive cycle there is an increase by $r_oN$. For the next reproductive cycle there are $N + r_oN$ individuals to multiply at the same rate. The change of trait strength, or population, from generation k to generation k + 1 is given by

$$N_{k+1} = (1 + r_o)N_k \tag{5-1}$$

This description subsumes a wide range of influences within the mean value of $r_o$. These include the random and cyclic effects of environment, predators, food supply, parasites, disease, etc. It also neglects such aspects as the variation of reproductivity with age (i.e., maturation) within the population and the effects of senescence and death in reducing the size of the reproducing population. Some accounting for these factors is implicit if $r_0$ is taken to be the overall population growth rate rather than the individual reproductive rate. We ignore these factors in the interest of simplicity and, as we will see, without too great a loss of generality. In the next section we will ameliorate this situation by including the effect of attrition.

As in the case of compound interest in banking, or the other examples given above, it is well known that Equation 5-1 leads to an exponential growth of N. To illustrate we start with $N_0 = 2$ and $r_0 = 1.5$, and obtain the values shown in Table 5-1. Note that a change in scale has

TABLE 5-1
EXPONENTIAL GROWTH

| k | N |
|---|---|
| 0 | 2 |
| 1 | 5 |
| 2 | 12.5 |
| 3 | 31.3 |
| 4 | 78.1 |
| 4.2 | 101.6 |
| 4.4 | 132.0 |
| 4.6 | 171.6 |
| 4.8 | 223.1 |
| 5 | 290.1 |
| 5.2 | 377.1 |
| 7 | 3998.9 |
| 7.2 | 5198.6 |
| 9 | 55128.2 |
| 10 | 204687 |
| 11 | 759989 |
| 12 | 2821786 |

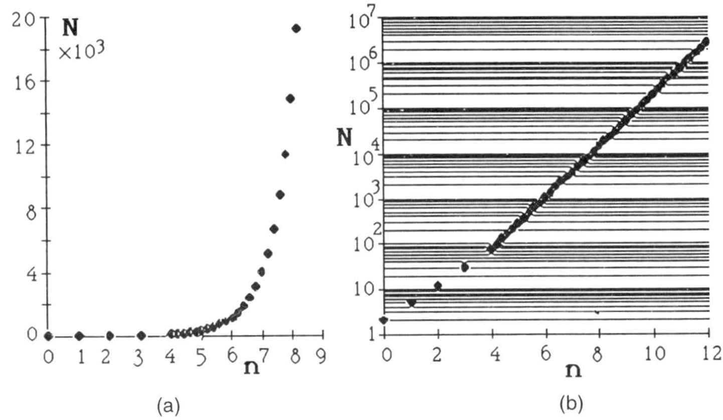


**Figure 5-1** Exponential growth. (a) Population number as a function of reproductive cycle number. (b) Logarithm of population number.

been made after generation k = 4. Beyond this point the population is large enough that we assume individual variability enters, so there is no longer a single, common reproduction cycle. This means that population growth becomes more uniform. To describe this numerically, we divide the cycle period into some number of parts, q, and assume that N/q members multiply during each subinterval. For a large population, one expects q to be large but for economy of calculation and space we have taken q = 5. (A mathematician would consider that this is equivalent to increasing the number of reproductive cycles by q and lowering the rate per cycle to $\mu = r_o/q$; this "effective" reproduction rate will have some significance in a later discussion.) From the table we see that in only 12 full cycles N achieves a value of nearly 3 million. Figure 5-1a is a plot of the values of $N_k$, showing its steep rise as the generation number advances. Figure 5-1b illustrates the value of the logarithmic representation of such rapid, exponentially growing numbers. Here the points fall on a straight line, demonstrating the specific exponential behavior and the property of unending growth.

## Logistics

However, unending growth cannot occur. Some interfering processes or inherent restrictions invariably impose an upper bound on N. For a population these may be the availability of food or the incidence of disease resulting from crowding, predation by other species, or internal fighting for space when N becomes too large. In the case of a growing, young animal it has been shown[1] that its ultimate physical size (in our context the ultimate cell number) is limited by the ability of its adult skeletal structure to bear the buckling forces due to weight and muscular shear stress. Let us consider two models that are commonly used to describe the imposition of a limit.

- Model 1: As the limit, $\eta$, is approached these restrictive pressures cause the total population regeneration rate to decrease from its uninhibited value $r = r_o$. For each reproductive cycle we take r to be given by

$$r = r_o[1 - (N_k / \eta)] \tag{5-2}$$

Note that well before the limit is approached, when $N_k$ is much smaller than $\eta$, the term $(N_k/\eta)$ is near zero so that $[1-(N_k/\eta)]$ is indistinguishable from 1 and we have $r \approx r_o$. When $N/\eta = 1/2$, i.e., half way to the maximum, we find that r has decreased to $r_o/2$. Finally, when N rises toward $\eta$ the fraction $(N_k/\eta)$ is nearly unity so $[1-(N_k/\eta)]$ becomes very small and the value of r approaches zero, meaning that the total population just maintains itself.

- Model 2: As the population becomes larger the death rate increases. This can be attributed to the same mechanisms mentioned two paragraphs above. In most wild animal populations, disease and predation are the limiting factors in longevity and population growth. This eliminates the need to consider the effects of senescence, discussed earlier. To describe this systematic effect we take the probability of the death of any individual to be proportional to the total population size, say $\beta N$. Then the total population death rate is $\beta N^2$, and Equation (5.1) becomes[2]

$$N_{k+1} = (1 + r_o)N_k - \beta N_k^2 \tag{5-3}$$

This is called Verhulst's law, after the Belgium sociologist who corrected Malthus' statement (equivalent to Equation 5-1) that human population would increase beyond the capability to maintain it. The effect of the additional term in Equation 5-3 is to limit the population to a value $\eta$ for which the average individual birth and death rates are equal: $\beta\eta = r_o$.

The two models reduce to the same growth relation. Combining Equations 5-1 and 5-2 leads to the alternative form

$$N_{k+1} - N_k = r_o N_k \left(1 - \frac{N_k}{\eta}\right) \tag{5-4}$$

We should note that these models involve only two parameters, $r_o$ and $\eta$ (The value of $N_o$ is not significant so long as it is small). Despite this simplicity, they have been successfully employed to describe a very wide range of disparate, complex phenomena,[3,4] including crop productivity, mining output, livestock prices, stock market fluctuations, insect appearances, epidemic frequencies, archeological diversity, etc. In part this versatility is because these equations are an example of the mathematical subject[5] of "Chaos." The deep and rich variety of behaviors that can be extracted from them is an extremely active and important area of research. We will discuss this later.

To illustrate the growth patterns of interest here, we take the figures used to derive Table 5-1, and a population limit of $\eta$ = 500,000 (or a per capita mortality rate $\beta = 3 \times 10^{-6}$). We need not give the values of N on a Table since the growth curve is shown by the open points in Figure 5-2. This curve has the sigmoid shape that is characteristic of limited growth; it is known as the logistic curve.

Following an initial, youthful period of exponential growth the logistic curve of Figure 5-2 has an adolescent, linear transition region, after which maturity is reached and growth ceases. The curve is symmetric about its midpoint. These behavioral features are common enough in developing systems that, as already mentioned, the simple logistic description has been remarkably successful in describing a wide range of phenomena.

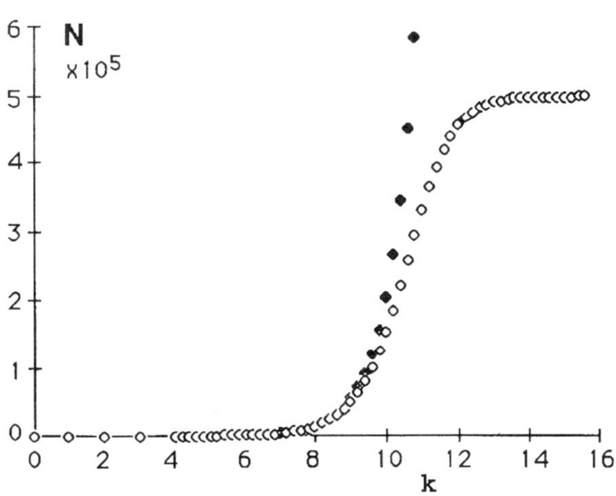


**Figure 5-2** Logistic curve for limited growth. Filled marks are from Figure 5-1. ($\mu = r_o/q = 0.3$)

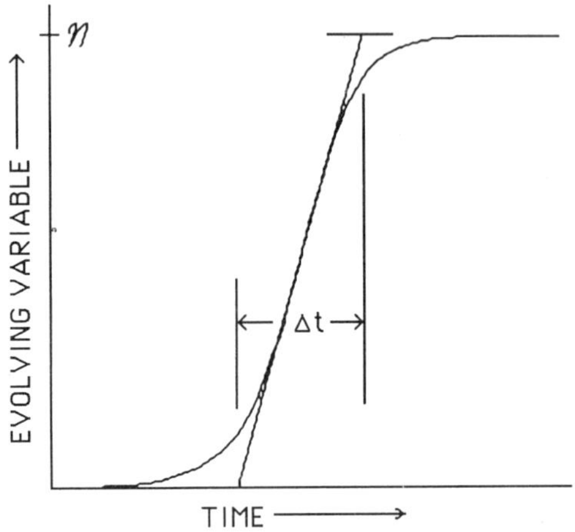


**Figure 5-3** Logistic curve showing the mid-point, tangent line used for estimating the length of the transition period, Δt.

Derek J. De Solla Price[3] has analyzed the features of such curves. He takes, as a measure of the transition time to saturation, the intercepts, with the upper and lower limits, of the line through the midpoint. This is shown in Figure 5-3. He estimates that the transition has a length equal to six times the period for population doubling of the initial exponential. It is shown in Appendix I that, according to our model, the transition interval is $\Delta t = 4/r_o$, where $\Delta$ denotes a change of t. $\Delta t$ equals 5.77 times that doubling time. As seen from Figure 5-3 this is a scant but adequate measure of the transition from youth to maturity. The exact multiple is unimportant but it is striking to note that the transition period depends only on $r_o$, the coefficient in Equation 5-4.

## Emergence

As already noted, from the known relationships presented in Chapters 2 through 4, or from nearly any other presentation of evolution of and on the Earth, one cannot escape awareness of

> ... that universal progress from evolution of elements to the appearance of conscious beings and of the human community, ...[6]

Such observations have been extant since the earliest attempts to classify the categories of nature's diversity. Until the mid-nineteenth century, that progression was regarded as representing God's fixed and immutable ordering of his universe, with mankind at its apex. In modern times, the indicated progress is usually taken to imply a dynamic continuity and a relatedness of the biological stages and participants of the evolutionary process. A more detailed and particularly perceptive summary of the common observation of this, and of the rate of progress of biologic evolution, was made by Professor William Day and is given in the frontispiece quotation.

We note that the above quotation counts the "human community" as an evolutionary stage, and the frontispiece infers that even this may only be temporary, in "the sequence of evolution's advance." The idea of a relation between biological and social evolution is not novel but we must be careful to distinguish its present form, as outlined in Chapter 4 and explicated in later chapters, from an older concept. Following the appearance of Darwin's theory, many social activists and philosophers attempted to use it to support their own ideas of development in current human affairs. Among other aspects this led to the promotion of racial superiority, national dominance, and Social Darwinism. This last, for example, proposed that societal structures could be altered to promote progress to higher, more advanced forms of government, economics, social intercourse, etc. While the impact of these ideas was considerable, some would say right up to the last decade of the twentieth century, they are largely based on misconceptions, incomplete analyses, and prejudices of interpretation, and are not germane to our study here.

The modern study of society, in an evolutionary context, has been from two different points of view. Some historians, anthropologists, and sociologists have treated the development of societal structures as direct outgrowths of the relations that evolved during mankinds evolution. Others have studied the development of individual institutions, customs, social structures, etc., as new forms of evolving systems of their own type, though analogous to the greater biological evolution. These efforts, and the quotations given earlier in this section reflect a qualitative understanding of the relation and occurrences of evolutionary progress, but suffer from not having a quantitative bases of comparison. It is our goal here to develop some of that quantitative perspective.

In our presentations in Chapters 2 through 4, some specific scenarios for relevant evolutionary changes were discussed, along with some specific means by which they might have occurred, particularly in dealing with the biological. These are certainly not indisputable and in nearly every case alternative explanations and mechanisms have been proposed. None of them, including the ones given here explain all the known facts and none are without at least some seemingly contradictory evidence. The particular descriptions presented either represent the consensus of experts or, where no strong consensus seems to exist, have been chosen on the basis of conformity and continuity.

However, regardless of the scenarios at each of these evolutionary changes, and probably at less prominent ones, a new stage of development has appeared. The changes that occur, e.g., between phylogenetic stages, do more than merely modify the older form from which they develop. Their resultant is of a fundamentally different nature, with the appearance of distinctive characteristics. Such observations have lead naturally to a theory of "Emergence," which postulates that in highly complex, aggregate systems, such as biological life forms or societal structures, a new configuration or form of behavior can be generated spontaneously.[7] It is widely assumed[8] that this is equally true of prebiological, organic, and postbiological changes.

Emergence is a holistic theory because the new whole is more than merely an extension of its precursors or than a rearrangement of its parts. The new properties and behaviors can not be predicted from only an analysis of the antecedents or of the isolated constituents. The need for different descriptions at different levels is not unique to evolving systems. Consider that the properties of a liquid or of a solid are more than a sum of the properties of the isolated atoms of which it is composed. The different descriptions and analyses that are needed for each range from the properties of free gaseous particles, through those of the condensed liquid, to the coherence of the crystal. Each has a unique positioning of the elemental components, and of their particular interactions and relations, that distinguish the greater systems from being merely the sum of its components.

In this regard, we note that symbiosis is also the intimate joining of dissimilar parts in a mutually enhancing relationship. It can be illustrated by the existence of intestinal bacteria, that are provided with food and a stable environment in exchange for which their action helps prepare that food for ingestion and absorption by the host animal. Another case is that of the lichens, plants that are partly algae and partly fungi. The two live in such intimacy that they appear to be a single entity. The algae are protected and supported by the fungi which, in turn, receives from the algae organic material it cannot produce itself. A third example might be the association of two complementary businesses, e.g., a printer and a binder, that form an association allowing them to seek projects that are inaccessible to or not cost effective for either alone. In these examples the partners are interdependent through their joining, which gives them versatility and adaptability neither would have alone. However, they remain discrete and their joint ability is little more than the cooperative sum of their individual actions.

Emergence goes beyond symbioses; it implies a much more powerful synergism of the joined elements. In synergism one is dealing with reinforcement and enhancement of existing properties to generate an additional degree of behavior or activity. With emergence, the new behavior exceeds mere enhancement of what existed before; it is another, distinct level of existence. It results from the unique reaggregation and altered interactions of modified elements. Although we cannot understand the new system, on its higher level, without understanding the history and components, that alone is not sufficient either to understand or to describe it.

It is important that we not confuse emergence with an older notion of a "vital force" of life.[9] This is the supposition that, since the earliest biological systems appeared, something beyond the chemical/physical body has been a part of organic systems, to make them live and develop. Although such an *elan vitale* has no basis other than the idea itself, it was once regarded as a progressive growing influence. Today it is acknowledged to be a regressive suggestion, based on a mystical response to a difficult question, and it has been rejected by nearly all students of life. Not only is it *ad hoc*, i.e. without evidential support, but it is nonproductive in that it leads to no new avenues of investigation.

## Higher Organization

The concept of emergence would, at first, appear to suffer from the same weaknesses as the vital force. However, it has recently been cast in a form that is more sophisticated, more scientifically satisfying, and more conducive to analysis. This is based on a "Principle of Higher Organization"[10] that takes as its starting point the hierarchy of organizational structures. Complex systems are conceived of in terms of several or many organizational levels in each of which the components are of a similar kind, in terms of physical and/or behavioral properties. These elements interact among themselves, they are subject to constraints and control imposed by higher levels of the hierarchical structure, and they experience limitations derived from properties of lower levels that they manipulate and upon which they depend.

As an example, consider what we would normally take to be the lowest level of a digital computer, the solid-state physical device that is a p–n junction. It is built on an understanding of the quantum mechanical laws of electron behavior in semiconductor materials. By design, according to the limitations and possibilities of these laws, this artificially constructed, microscopic structure can act as an electrical switch. It is placed in an electronic circuit, constituting the next hierarchical level, that supplies electric signals to control the switching behavior. Conversely, in practice the effectiveness of these signals is limited in magnitude, speed, type, etc. by the capabilities of the junctions. On a higher level the electrical circuit is part of a system that directs the switching events, controls signals between circuits and devices, maintains stability, etc. Of course, the behavior of this system is limited by the possible circuit responses. On an even higher level, we have software instructions that direct the overall computer logic and convert the lower function to meaningful processes. Even beyond this we have interconnected computer networks that coordinate the software actions, etc.

We have pointed out that each stratum of operation of such a multilevel system, whether electrical or organic in nature, has its own rules and modes of operation. Although these remain in force at that level, their actions are modified insofar as they are governed, or controlled, or limited by the rules of the other levels. Generally, interactions have somewhat of a reciprocal nature, with the higher levels being constrained by the capabilities, forces, and rules of the lower levels. The successful existence and operation of such a far reaching network of causes and effects demands a stability that resists disruptions of its organization, that is, it limits the range and type of any new behavior that can appear. As a result, changes tend to be incremental; widely different types, actions, or forms cannot suddenly emerge from existing states.

What makes the evolutionary Theory of Hierarchical Systems novel, useful, and that gives it the potential to analyze emergent, evolving systems, are the exceptions it anticipates to this stasis. It is proposed that systems with many hierarchical levels, and particularly those whose elements have great complexity, can, under conditions to be discussed below, undergo spontaneous and sudden change. By this is meant that, if the system becomes sufficiently complex in its structure and interactions, a threshold may be reached beyond which a new mode of organizational behavior can appear. This would constitute an emergent change in the sense we have described; such a change is called a Metasystem transition.[11]

We should be aware that the "sudden" appearance of an emergent change may only be a relative rapidity. Over the geologic periods we are considering, punctuation of the existing equilibrium of biological systems may very well occur over millennia of time and still appear, archaeologically, abrupt. Even in the most recent case, of the development of electronic computation, its emergence can be seen to have taken place over several decades of time. However, systems analysts who apply these concepts to the structures of society, such as corporate or government agency histories, point out that, while the same considerations apply, the intervention of the human mind has resulted in the possibility, and in fact the realization, of almost instantaneous system changes.

The emergent changes we are discussing here appear to be beyond the logistic growth process described earlier in this chapter. However, this is not entirely true. Although we will present a more powerful extension of logistic growth in the next chapter, it is important to understand that a wide variety of such behavioral changes are described by variations of the logistic process given earlier. From empirical studies of logistic growth curves Derek de Solla Price[3] has pointed out that, as the midpoint is passed, it is not uncommon for the curve to depart from its expected, S-shape, altering the approach to senility. Several different departure modes are possible, and some of them can be derived simply from our mathematical model. For cases of vigorous growth (large $r_o$) and where the individual reproductive cycles remain more-or-less coherent (small q, e.g., if constrained by seasonal or annual cycles), so that $r_o/q$ is greater than unity, Figure 5-4a and 5-4b show the convergent

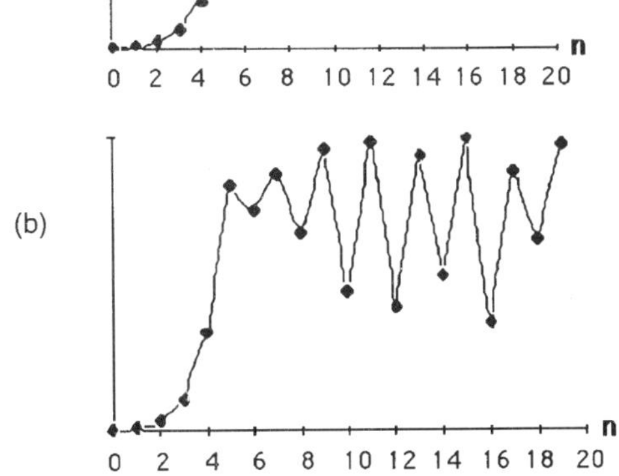


**Figure 5-4** Population dynamics. (a) $\mu = r_o/q = 1.75$, (b) $\mu = r_o/q = 2.6$.

and chaotic oscillation that can result.[5] We will see in the next sections that the descriptions of a range of population dynamics, involving different cyclic variations, can also be deducted from the properties of the same mathematical relations.[5]

## Deterministic Chaos

The true basis for the expectation of emergent changes in complex systems derives from the mathematical theory of "Deterministically Chaotic" systems, combined with the physical theory of nonequilibrium thermodynamics. Chaotic functions are a class of "non-linear" mathematical equations. By this is meant that, in the behaviors they describe, the response is not simply proportional to the cause. We normally expect proportional, or linear, response: the harder we pull on a rubber band the more it stretches, the louder we yell the further we can be heard, the heavier a support the greater its load bearing ability, and so forth. However, even for the simple behavior described by equation 5-1 the increase of N is more rapid than the uniform time intervals, and Equation 5-4 and Figure 5-2 change their direction and even stop increasing as their stimulus (the time) changes uniformly.

We can examine the range of behavior of Equation 5-4 by systematically taking some examples with different reproductive rates. For this it is convenient and appropriate to use the parameter we defined earlier: $\mu = r_o/q$, the effective reproductive rate per reduced cycle.

- Figure 5-2 shows the S shaped curve that results when $\mu < 1$ (on that figure $\mu = 0.3$)
- For $1 < \mu < 2$ we find the oscillating behavior illustrated in Figure 5-4a, that converges to a final value. In this and the case that $\mu < 1$ the population ultimately arrives at a single value.
- For $2 < \mu < 2.4495$ (equal to $\sqrt{6}$) it is found that the population number oscillates between two stable values, as shown on Figure 5-5a.
- Further increase of $\mu$ shows a narrow range where there are four stable states, causing the population to oscillate with a period of 4 transitions. See Figure 5-5b.
- Increasingly narrower intervals with periods of 8, then 16, etc. then appear as $\mu$ increases further.
- Beyond $\mu = 2.61547$ the population varies in what appears to be a truly random, i.e., chaotic manner. However, even here, the logistic equation gives a definite value at each successive iteration; the "random" values are, nonetheless determined by the equation. These features gave rise to the name of this exciting field of mathematical

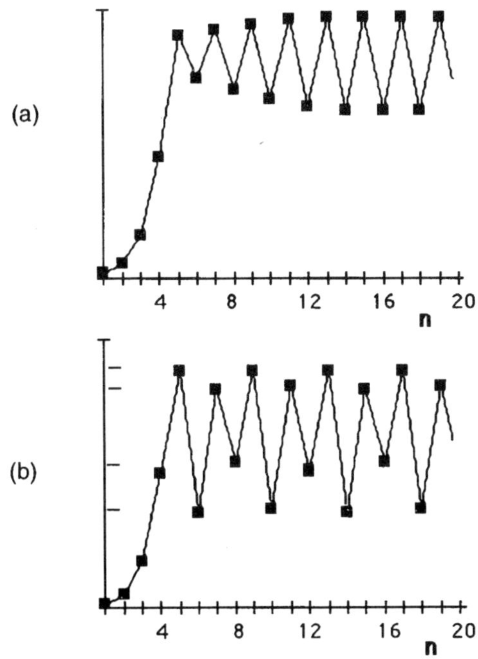


**Figure 5-5** Periods of Chaos function. (a) $\mu = 2.2$ (b) $\mu = 2.55$.

analysis: Deterministic Chaos. Figure 5-4b illustrates the behavior just below the onset of true chaos, where N appears to vary randomly but, for this value of $\mu$, still has a limit.

Each of the values of $\mu$, where the behavior changes, is called a bifurcation point, indicating that two modes of response are possible. At $\mu = 2$ it goes from one to two cycles, at $\mu = 2.4495$ it changes to 4 cycles, and so forth. These transitions are given on Figure 5-6 showing the ratio of the different population numbers to the maximum (the horizontal line at unity) at each $\mu$ value. This bifurcation diagram also indicates the number of levels at each value of $\mu$. For example, for $0 < \mu < 2$ there is only a single equilibrium state; for $\mu = 2.5$ there are four turning point values. The dashed line on the figure marks the onset of chaos, i.e., random variations.

The basic origin of this behavior is the non-linear nature of Equation 5-3. The non-linearity arises from the $N^2$ term, and this is, perhaps, the simplest non-linear equation. In the theory of hierarchical systems, it is assumed that many of the subsystems, at any organizational level, are non-linear in a deeper manner, and interact between themselves and between other levels in a non-linear way. As a result, change of behavior

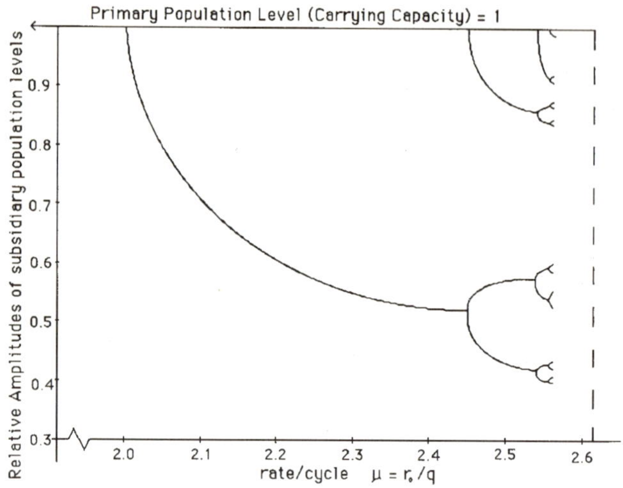


**Figure 5-6** Bifurcation diagram for the quadratic Chaos (Logistic) function. For explanation see the text.

at local and general bifurcation points can always be expected. If such a change or changes occur that are extensive and that produce profound alteration in the overall system, one would have a system transition, corresponding to an evolutionary emergent event. Systems that undergo such changes and rearrangements spontaneously are said to be self-organizing.

Other analyses and examples have been given[10] to support this principle of emergent self-organization,[11,12] and many have taken it as the basis for evolutionary change. The mechanisms by which such a change can take place (i.e., why $\mu$ would change its value) is described by the theory of physical systems that are far from their equilibrium state. We will discuss this in a later chapter.

## Synergetics

We should be aware of a challenge that is made to this ad hoc postulation of truly new, spontaneously generated behavior. It arises from the fact that the proposed emergence at a bifurcation point appears to violate the cutting rule of Occam's razor. William of Occam, an early fourteenth century scholar, was greatly bothered by the then common custom of proposing specific mechanisms, particularly theological ones, to explain each observed behavior.[13] This procedure destroys any expectation to finding a unity of mechanisms. He therefore proposed the guideline that "causes not be increased unnecessarily." Today this attitude is reflected in the universally accepted scientific dictum that no new laws or causes are acceptable in explaining an observation unless all others are exhausted and there is compelling reason and evidence to do so. (For this effort, and for his opposition to other Papal theological doctrines, William of Occam was convicted of heresy and imprisoned.)

In part because of Occam's principle, many efforts are being made to clarify and understand the emergence of hierarchical reorganizations and complexity changes in terms of more fundamental properties.[14] One such program is the exciting field of synergetics. "Synergetic systems are those that can produce macroscopic spatial, temporal or functional structure in a self-organizing way".[15] By self-organizing is meant the property or ability of a system to spontaneously, i.e. with no specific, outside ordering influence, alter its structure, function, or the nature of its cooperative interactions. Of course such reorganization is likely induced by general forces, such as departures from equilibrium, but these do not impose a specific form to the change that may occur. In Chapter 9 we will examine an example to clarify this difference.

Synergetics involves a search for mathematical models of self-organization, based on the cooperative interactions of the parts of a population, or on the dynamics that result from being far from equilibrium. Beside the interactions, the critical feature of synergetic models is their non-linear behavior. As already mentioned, in non-linear systems a response is not simply proportional to its stimulus. It may be more, or less, or directed differently. The generating causes may be random influences, such as noise and fluctuations, or changes of environment, or the drift through a variety of non-equilibrium states. These create initial changes and in a non-linear system such changes can grow. If the system passes through a bifurcation point, beyond which it has a different behavioral mode, this can result in a phase change, in divergent behavior, in a new order of component coordination, or in approaching a new equilibrium, i.e. in a system transition.

To illustrate some features of this analysis and its consequences, let us take a simple numerical example. The following is not intended to represent the true complexity of a hierarchical system but merely to indicate the difficulties that are encountered in performing and interpreting such quantitative studies, even in a grossly oversimplified case. Imagine a system described by

$$V = x + y \qquad (5\text{-}5)$$

where x and y might represent the components of one hierarchical level and V is its contribution to the next level. Initially x and y have the values x = 10, y = 20, so that V = 30. x contains a stabilizing factor that tends to maintain the value of V; its stability coefficient is $\alpha = 0.01$. The exact expression for x is shown in Equation 5-6.

$$x = 10\,[1 - \alpha(V - 30)], \qquad \alpha = 0.01 \qquad (5\text{-}6)$$

The feature of interest is that if V departs from 30, for any reason, the value of x changes slightly in the direction to restore V. $\alpha$ represents a weak reverse communication between the hierarchical levels, i.e., a negative feedback.

As shown in Equation 5-7,

$$y = 20\,(1 + \beta|\Delta x/\Delta t|), \qquad \beta = 0.055 \qquad (5\text{-}7)$$

y opposes this tendency of x through a dynamic sensitivity term with coefficient $\beta = 0.055$. The change of y depends on the rate of change of x (in Equation 5-7 it is actually proportional to the magnitude of this change). This represents an intralevel communication, an interaction between x and y.

We now suppose that, either through an internal change, such as a mutation, an infection, or an environmental change, a new interaction is established so that x also becomes directly sensitive to y. Instead of Equation 5-6, we now have Equation 5-8.

$$x = 10\,[1 + 0.1y - \alpha\,(V - 30)], \qquad \alpha = 0.01 \qquad (5\text{-}8)$$

When this happens the value of x changes immediately and, through Equation 5-7, so then does y. These changes alter V which, in turn, alters x, initiating a second cycle; the process continues indefinitely. We can observe the resulting variations by taking time changes in uniform steps; the behavior of the system value function, V, is shown in Figure 5-7(a). After an initial increase and oscillation it decays to a new value, V = 50. (That this value is different from 30 implies that x and y must be oscillating in the new state so that, although it is not apparent from the figure, the new equilibrium is a dynamic one.)

Behavioral sensitivity to values of the interaction coefficients, $\alpha$ and $\beta$, can be demonstrated if we change $\beta$ very slightly, to $\beta = 0.06$. Using

Equations (5),(7) and (8), Figure 5-7(b) shows that V now diverges, oscillating between increasing limits. Another feature, found in some non-linear systems, can be illustrated by this example, the existence of "local attractors." These are sets, or loci, of parameter values, or relations between those values, toward which the system tends to move, though it may never arrive exactly there, or between which it switches. In the case of Figure 5-7b, it is found that while x, y, and V increase after the change of $\beta$, the system continually cycles between states for which the ratio x/y is 0.791336, then 23.1642, and then –0.04802.

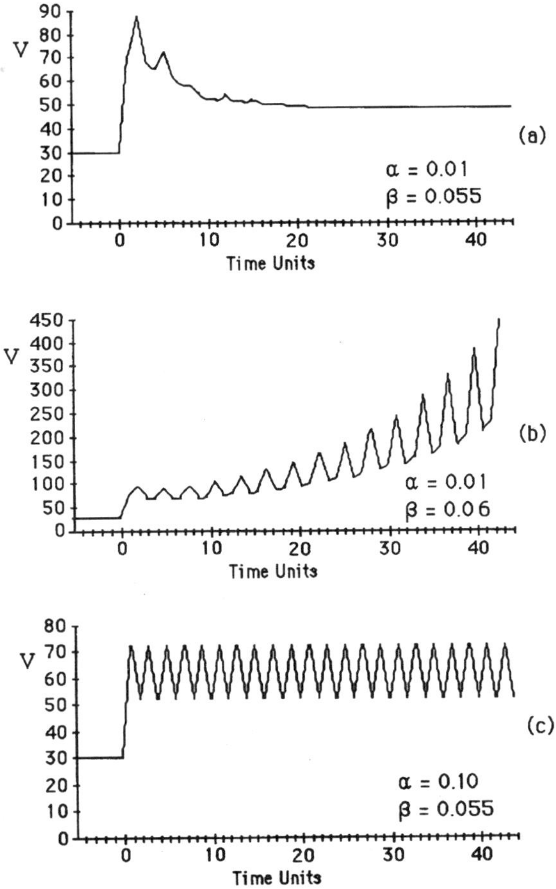


**Figure 5-7** The effect of parameter changes in a Synergetic system.

In this example, the effect of greater stabilization can be seen by restoring $\beta$ to its initial value of 0.055 and increasing $\alpha$ to 0.1, yielding the regular oscillations shown in Figure 5-7c.

The important lesson of these results is to demonstrate the inherent need for mathematical relations in describing synergetics, and to indicate the wide range of outcomes that can be yielded by minor variations of synergetic models. This last point is a particularly important property of non-linear systems. Many of these display a great sensitivity of the long term behavior to the exact starting value of the variable or variables being studied. By this is meant that very small initial differences can result in completely different ultimate behaviors and values. Specific results are therefore difficult to interpret and to generalize. Whether any behavior, such as those illustrated here, should be considered to represent a different state of order would also have to be demonstrated for each case. The variables to be used in each model, here the V, x, y, need defining from real system. This example is very simple, having only three subsystem: V, x and y. Real systems, particularly with large hierarchical structures, are vastly more complex in terms of the number of elements at each hierarchical level, in the number of levels, and in the number, variety, and variability of interactions. At least in organic systems, complexity itself is generally taken to be the appropriate evolutionary value function in theories of higher organization. The concept of complexity is not well defined though attempts are being made to do so. However, it has an appealing reasonableness and recognition.

# 6: CYBERNETIC EVOLUTION

## Logistic Escalation

In the previous chapter we examined the logistic equation and the sigmoid growth curve it describes. In his study of logistic behaviors Derek de Solla Price[1] pointed out that, after reaching saturation, one frequently sees a behavior called "escalation."

> If a slight change of definition of the thing that is being measured can be so allowed as to count a new phenomenon on equal terms with the old, [a] new logistic curve rises phoenix-like on the ashes of the old ...

He discusses several examples of escalation. These include the energy of particle accelerators, the number of universities, and the number of known elements. This last case is shown in Figure 6-1, which is an updated version of Price's presentation (to 1996).

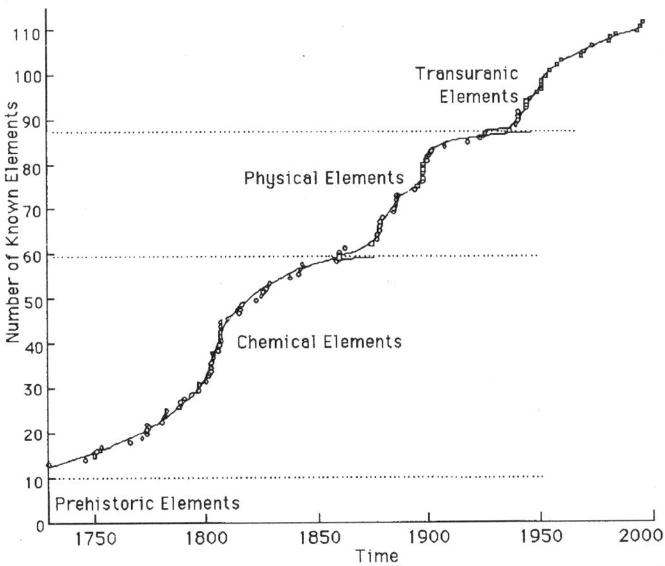


**Figure 6-1** Logistic escalation of number of known elements.

Ignoring the few elements known from antiquity, the figure shows a first period of discovery involving a logistic growth of the known number through chemical discovery, between 1700 and 1850. When chemical differences of the elements had been exhausted, the resulting saturation is ended by a second escalation, lasting until about 1940. Here physical means, such as mass spectroscopy, optical spectra, and nuclear identification were used to separate chemically similar elements. This stage ended when all the natural elements had been discovered and the third escalation occurs due to the creation of transuranic, manmade elements in high energy machines. Although this search continues, with element 112 having been recorded in early 1996,[2] the S-shaped ogive appears to have nearly saturated as a result of limits on the stability of these nuclei. The higher elements are so unstable that only a few atoms have been created and these have lifetimes of milliseconds, or less, and merely identifying them is extremely difficult. Modis[3] has described these changes with four cycles, though neither recognizing nor interpreting them specifically as emergent changes.

Price's "change of definition of the thing that is being measured" implies the prominence of a new aspect of that growth index. One certainly expects that any parameter of a complex system has may dimensions, some subset of which may be activated under different conditions and at different times, leading to cycles of escalation. The simple form of the logistic equation, given in Chapter 5, does not describe this behavior but the logistic description can be modified to include the phenomenology of escalation. For this we have the following (for the four cycle case):

$$N_{k+1} - N_k = r_o N_k \left|1 - \frac{N_k}{\eta_1}\right| \left|1 - \frac{N_k}{\eta_2}\right| \left|1 - \frac{N_k}{\eta_3}\right| \left(1 - \frac{N_k}{\eta_4}\right) \quad (6\text{-}1)$$

To understand how this equation describes the desired properties consider that the simple logistic equation for $N_{k+1} - N_k$, Equation 5-4, is a product of two terms. One is proportional to the departure of N from its starting level ($N \approx 0$) and one is proportional to its separation from the saturation level: $(1 - N/\eta)$. For the first escalation in Equation 6-1 these are also the first two terms; for the second escalation we shift one factor to the terms involving $\eta_1$ and $\eta_2$; for the third escalation the terms in $\eta_2$ and $\eta_3$ serve the purpose, and so forth for as many escalations as occur. The vertical bars on the middle factors are "absolute value" signs, meaning that these terms are always positive. As a result, when N is very close to one of the $\eta$, i.e., near saturation at that value, a small perturbation will cause it to begin a new growth cycle toward the next $\eta$ limit. (Only when approaching the final $\eta$, here $\eta_4$, does a negative sign occurs if $N_k > \eta_{final}$, causing the series of escalations to terminate.)

During the first escalation, when $0 < N < \eta_1$, the higher numbered factors, with $\eta_2$, $\eta_3$, etc., have not yet been realized and are therefore ignored. As a result the single logistic form ensues, as in Chapter 5. In the next escalation, between $\eta_1$ and $\eta_2$, the even higher factors, with $\eta_3$, $\eta_4$, etc., are still unity but now a memory of the first cycle remains because the rate constant, $r_o$, is replaced by $r_o^1 = r_o N_k/\eta_1$. This rate constant is slowly varying but, more importantly, it is larger than that of the previous cycle, i.e., $r_o^1 > r_o$. This same reasoning applies to each higher escalation since, after each transition the rate constant increases by an additional factor. As a result, $r_o^j > r_o^{j-1} > \ldots r_o^1 > r_o$. Since we saw that the transition times are proportional to the 1/r, it follows that consecutive cycle times decrease.

By means of a more careful mathematical analysis it is shown in Appendix II that the ratio of two successive escalation transition times, from initiation to saturation of the cycles, is given by

$$\Delta t_{k+1}/\Delta t_k \propto R_k^{-k}, \quad \text{where} \quad R_k = \eta_k/\eta_{k-1}. \qquad (6\text{-}2)$$

$R_k > 1$ so that the series of escalations converge, that is, the successive time intervals become smaller and the escalations approach a limit. If the R change slowly with cycle number then the appendix also shows that the total time from the end time of the sequence of escalations, back to the k-th transition $t_k$, is

$$\log t_{k\,(past)} = -k \log R + \log C \qquad (6\text{-}3)$$

On a plot of $\log t_k$ versus k, this corresponds to a straight line, called the trajectory of the particular evolution process.

This behavior is a characteristic of escalation sequences described by equations such as 6-1 and it follows that the display of such a straight line demonstrates that escalated logistics is the means of emergence of new stages of any evolution being studied. We now present the semilogarithmic plots for several widely different types of evolution. This demonstrates that, for these phenomena, escalated logistics is the means for the emergence of new stages.

## Examples

1) *The Physical Domain*

Having the data in Figure 6-1 for escalations of the discovery of the chemical elements, we can test Equation 6-3 directly. This is shown in

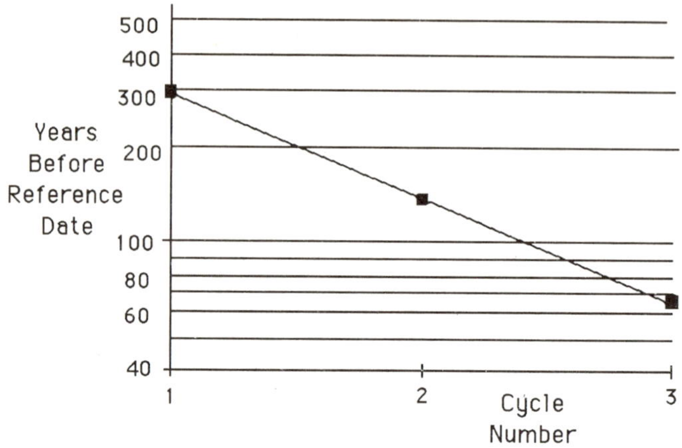


**Figure 6-2** Trajectory of increase in the number of known elements.

Figure 6-2. In examining the apparent good agreement of this figure with the expected behavior of Equation 6-3 we must be aware that there is uncertainty in selection of the dates from Figure 6-1, as well as in the determination of $\eta_k$, where the logistic cycles begin and end. These errors are relatively small and do not greatly alter the straight line shown. However, the evidence provided by only three points is barely convincing, especially since errors and differences are reduced by taking the logarithms of the point values.

2) *The Biological Domain*

In Chapter 1 we saw that those transitions marking changes in the dominant life form on earth also conform to a semilog-linear plot (Figure 1-3). In view of our understanding of hierarchical structure we interpret that figure as a relation for the development of taxonomic complexity. Overall biological evolution therefore appears to represent a sequence of primary, emergent escalations followed by intervals of more gradual change, during which individual species may experience their own secondary periods of stasis and punctuated change. The least-squares fit of those points to the equation of a straight line:

$$\log t_{k(\text{past})} = -mk + b$$

gives $m = 0.502 \pm 0.062, \qquad b = 4.28 \pm 0.11.$

The trajectory of taxonomic complexity seems to have terminated before reaching event X. This observation is already broadly appreciated; in discussing "The Dampening of Evolution," Pierre-P. Grassé[4] notes that

> From the facts already discussed, one notices that the "Maneuvering space" of evolution has never stopped decreasing. The genesis of the phyla stopped in the Ordovician; of the classes in the Jurassic; of the orders, in the Paleocene-Eocene. After the Eocene, the evolutionary "sap" still flowed through a few orders, since mammals and birds continued to specialize in various directions and invaded all the territorial and marine biotypes occupied by reptiles.
> The extent of evolutionary novelties gradually changed. They no longer affected the structural plan but only involved details. The only form which evolution took was speciation: in insects since the Oligocene, in mollusks since the Miocene, in birds and simians since the Pliocene, and in some grilines and hominids since the Holocene; *Homo sapiens*, the last in line is probably 100,000 years old.

We will be able to shed some illumination on the relation of this termination to other evolutionary trends after we have developed an alternative representation of evolution on earth.

3) *The Social Domain*

Emergence of each escalation of an evolving system involves a change of the system growth parameter, as it nears its previous maturation. This represents a fundamental change of the system, i.e., the first appearance of a new feature or features, or of a new behavior. As we saw, in example 2, above, it is frequently easier to recognize these critical changes, or the transition events that indicate their appearance, than it is to define the evolving parameter itself. In such cases it is possible that continuity of the logt vs k trajectory can be a useful tool, not only for examining the parameters of evolution, but it may even indicate events that would be overlooked otherwise.

Consider the well-known changes of monotheistic religious awareness and belief, in the Mideast and Europe. Although there has surely been an evolution, it is rarely regarded as a single, continuous, progressive process, in the same way as biological evolution or even as the discovery of elements. Yet, it has been marked by a series of occurrences that changed its form in distinctive ways and that can be considered transition events. They are distinguished by the names of the powerful, charismatic individuals who initiated them; there can be little disagreement that pivotal, historic changes were wrought by those indicated in Table 6-1.

TABLE 6-1
PIVOTAL RELIGIOUS EVENTS

| EVENT NUMBER | HISTORIC INDIVIDUAL | YEARS BEFORE PRESENT[a] |
|---|---|---|
| 1 | Abraham | 3800 |
| 2 | Jesus | 2000 |
| 3 | Muhammad | 1400 |
| 4 | | |
| 5 | Luther | 500 |

[a] with respect to the year 2000

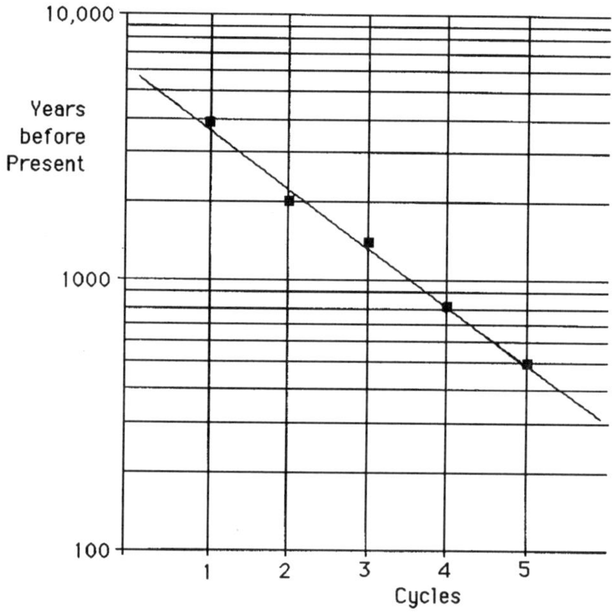


**Figure 6-3** Trajectory of events of the logistics of religious awareness.

Note that row 4 has been left empty. This was done because the semilogarithmic trajectory of points on the table, shown in Figure 6-3, does not constitute a straight line unless the Lutheran change is translated from position 4 to position 5. The missing event, to fill position 4, falls 800 years ago. The events on Table 6-1 affected the lives and beliefs of millions of people, causing conflicts in which tens and hundreds of thousands were killed, and their influence have been felt through the centuries. For an event of this magnitude a literature search quickly discloses that this is the time of the Crusades. Relevant quotations by scholars of this historical period reveal that:

> The era of the Crusades is one of the most important in the history of Western civilization.[5]

and

> ... the Crusades were the outcome of an enthusiasm more deep and enduring than any other that the world has witnessed.[6]

and particularly

> The fourth Crusade culminated in the sack of Constantinople by the Crusaders and Venetians in 1204. Few episodes in European history have had so great and so far-reaching an effect on western civilization.[7]

In view of this it is appropriate to fill row 4 of the table with the following:

| | | |
|---|---|---|
| 4 | Crusades | 800 |

thus completing the points of the figure.

The form of presentation of Figure 6-3, and the inherent open endedness of its subject leads one to extrapolate the trajectory shown to a preceding point approximately 6000 years ago. We may note that it was at about this time that the earlier shamans of primitive societies were transformed into priests of the more mature statehoods of mideastern history, with the implications that change had on the mode and meaning of worship and on the nature of the gods. However this is, perhaps, not consistent with the other, more well defined changes and we should be cautious about this assumption. A forward extension seems more likely and would lead us to expect a transition at about 300 BP, the year 1700 of our calendar.

There were several events of religious and political significance surrounding this date, but none produced as profound a change as those on Table 6-1. These include founding of the Society of Friends (ca 1650), the Jewish upheaval associated with Shabbetai Tzevi (1665), and, somewhat earlier, the Puritan/Pilgrim founding and emigration (1600–1620), and, later, founding of the Mormon Church (1820–1830). We might therefore seek a reason for interruption of the trajectory at that point. For this we turn to another progression that paralleled the religious evolution.

In the year 1687, Isaac Newton published his epoch making *Principia*, describing how the laws of gravitation control the motions of all bodies, both those on earth and those of the cosmos. To establish his hypothesis, Newton developed the powerful tools of the differential and integral calculus, giving his work the power of irrefutability. Historians of science[8] have called this "... perhaps the greatest event in the history of science." Beside revolutionizing the concept and conduct of physical science, it also initiated a revolution of religious thought by raising the question: If all bodies, large and small and including even the heavens, are controlled by impersonal mathematics then what is God's role?

If Newton's *Principia* raised questions of the immanence of God, then Darwin's work on evolution, published in 1852, with its denial of the special creation of mankind, was devastating. Even more than Newton's work this seemed to remove God from our history and to substitute a scientific rationale. These works clearly constituted critical transitions in regard to understanding the mechanisms of the world; they were steps in the shift of "world-view" from the purely theological to one whose basis is scientific and technological.

Of course the process did not end there. Early in the twentieth century, the physical sciences made several quantum advances that resounded in the public mind. One was Albert Einstein's theory of Relativity (1905) and a second was the fundamental development of quantum theory itself, between 1915 and 1925. James Clerk Maxwell presented his theory of electromagnetism in 1864 but full advantage could not be taken of it until the invention, in 1907, of the vacuum tube triode by Lee DeForest. This was the device needed for radio communication to become a public phenomenon. Almost simultaneously, the initiation of manned flight by the Wright brothers (1905) also captivated the public mind. These events indicated both the progress of science and the emergence of engineering as a major factor in the awareness of that progress. Together we take these theoretical and engineering advances to define an evolutionary transition at about 85 BP.

Closer to the present, the mid-twentieth century has witnessed a recent critical change attributable to engineering, that has altered the ways in which science is conducted, in which society functions, in which we communicate, and in which we view our world. Spurred by advances of the preceding 30 years and supported as part of the program of the Second

TABLE 6-2
CRITICAL CHANGES OF WORLD-VIEW

| | religion | science | event |
|---|---|---|---|
| 1 | 3800 | | Abraham |
| 2 | 2000 | | Jesus |
| 3 | 1400 | | Muhammad |
| 4 | 800 | | Crusades |
| 5 | 500 | | Luther |
| 6 | | 313 | Newton |
| 7 | | 148 | Darwin |
| 8 | | 85 | * |
| 9 | | 55 | Computer |

* DeForest, Wright Bros., Einstein, Schrodinger, Freud

World War, the inventions of the digital computer (1943) and of the transistor (1947) have lead to our present "information age." Though "digital electronics" probably best describes this transition, we denote it by its more common reference as the advent of the computer.

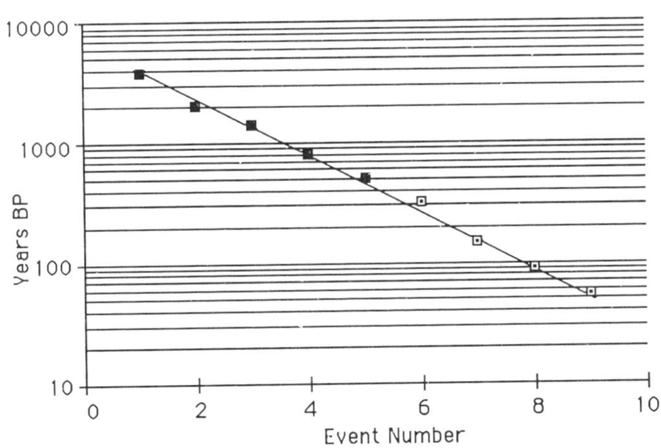


**Figure 6-4** Trajectory of the change of World-View in 'Western' society.

Adding these scientific events to those of Table 6-1 we arrive at Table 6-2, showing the combined critical changes of world-view. Figure 6-4 is the trajectory of these points. It demonstrates that the trajectory of religious change constitutes only the first phase of the systematic variation of world-view of Western society. This did not so much terminate in the eighteenth century as the enthusiasm it engenders was blunted and redirected along less supernatural lines. The meaning of this change will be briefly discussed later.

## The Evolutionary Trajectory

In the previous section we discussed the trajectory of changes of dominant life forms displayed on Figure 1-3; it represents taxonomic — morphological evolution. As acknowledged, for example, by the quotation given there from Pierre-P. Grasse, and as discussed in Chapter 3, this trajectory of evolution appears to have terminated some 100,000 BP, with the appearance of *Homo sapiens*. However, our discussion in Chapter 4 clearly shows that developments of a different sort have continued through human effort. Manfred Eigen[9] puts this in perspective when he states that

> The process of creation is by no means at an end, although no-one can predict what is to come, even within intervals of time that are negligibly short in comparison with the phase of genetic evolution.... But evolutionary progress in the near future will hardly be on the genetic level. The activation of the human mind has greatly speeded up the roundabout of development. Almost everything that happens in the foreseeable future will proceed from mankind.

We therefore seek an evolutionary trajectory that is more encompassing than that shown in Figure 1-3. In such an undertaking the selection of critical transitions is certainly not obvious so we must be careful to refer our choices to the standards, criteria, and divisions established by expert scholars in their respective fields. For this reason, we divide our presentation into three series of events: preceding, during, and following organic evolution. In each of these domains we will find that there are well-defined stages of development and acknowledged transitions.

- PRE-EUCARYOTIC EVOLUTION. From our discussion in Chapter 2 we, along with others,[10,11] take the events on Table 6-3 to be the critical occurrences of pre-eucaryotic evolution. All subsequent evolution is contingent on them.

- ORGANIC EVOLUTION. From even before Darwin's publications, nearly 150 years ago, and certainly since then, the key events of organic

TABLE 6-3
EVENTS OF PRE-EUCARYOTIC EVOLUTION

| NUMBER | EVENT | DATE (BP) |
|---|---|---|
| 0 | Big Bang | 1.0–2.0 $(10)^{10}$ |

Being the origin of all we know and have, this must certainly be considered a critical point. It is *the* beginning — the origin, and we therefore denote it as zero.

| NUMBER | EVENT | DATE (BP) |
|---|---|---|
| 1 | ⌈Solidification of Earth⌉ | 3.6 $(10)^9$ |
| | ⌊First Life (Procaryotes)⌋ | 3.4 $(10)^9$ |

On the scale involving $10^9$ years these events occur in such close proximity that we take them together. The significance of this will be discussed in chapter 10.

| NUMBER | EVENT | DATE (BP) |
|---|---|---|
| 2 | Eucaryotic radiation | 5–8 $(10)^8$ |

After the very long intermediate development of reproducing molecules, simple cells, and tissue and organ evolution, there was an explosive spread of relatively sophisticated life forms. Event 2 marks the beginning of metazoan evolution, i.e., of multicellular animals.

evolution have been deeply considered and debated by archeologists, paleontologists, and biologists. In spite of this, or perhaps because of it, the selection of a set of critical events is not clear; in part it depends on the criteria adopted. Therefore note that:

- As it reflects the acknowledged cladistic history and the existing and fossil physical reality, an appropriate basis for selection is the normal, well established, phylogenetic and taxonomic system of divisions into which biologists and paleontologists divide living things, past and present.
- We do not know what is the correct evolutionary parameter or parameters that guide the trajectory we are constructing. Therefore we should be prepared to find that the usual biological criteria may not be entirely or exclusively appropriate. In fact, we will later see that, even during these stages of organic change we are dealing with a growth of information.
- Although many object to the anthropocentric view of organic evolution we must recognize, from all our previous considerations, that *H. s. sapiens* has come the farthest from the origin of all life, $3.4(10)^9$ years ago. Furthermore, as we have already noted, it is widely acknowledged that the activities of humankind have advanced evolution beyond the organic domain.

Therefore, in considering organic evolution we will use the standard, textbook phylogeny of *H. s. sapiens*. Note that we do **not** propose an evolutionary scaling that can be fit by all species but, rather, the transitions of human evolution are uniquely indicative of the greatest progress of evolution, and it is this that we seek.

Table 6-4 is extracted from Chapter 3, showing a typical listing of the phylogeny of mankind, along with the dates of the earliest appearance of

TABLE 6-4
PHYLOGENY OF MANKIND

| NUMBER | CATEGORY | LINE OF DESCENT | DATE (BP) |
|---|---|---|---|
| 3 | CLASS: | Mammal | $1.5–2\ (10)^8$ |
| This marks the appearance of a sophisticated, versatile lifeform that cares for its young and that can thrive under a wide range of conditions. | | | |
| $\alpha$ | ⌈ORDER<br>⌊SUBORDER | Primate<br>Anthropoid | $\sim 7\ (10)^7$ |
| Note that the Order: Primate and the Suborder: Anthropoid are bracketed together and not assigned a sequential number. This will be explained shortly. | | | |
| 4 | SUPERFAMILY | Hominoid | $2.5–4.0\ (10)^7$ |
| The living Hominoid apes are gibbons, orangutans, chimpanzees, and gorillas. These are in parallel families to the line of descent of humanity and its immediate ancestors. | | | |
| 5 | FAMILY | Hominid | $4–10\ (10)^6$ |
| Creatures that resemble man. | | | |
| 6 | GENUS | Homo | $1.5–2\ (10)^6$ |
| The first appearance of this Genus signifies a change of evolutionary emphasis from physical and anatomical change to the attainment of greater brain size. | | | |
| 7 | SPECIES | H. sapiens | $2–5\ (10)^5$ |
| Transitional Homo Sapiens have many of the attributes of modern H. Sapiens but are not fully evolved. | | | |
| 8 | SUBSPECIES | H. s. sapiens | $7–10\ (10)^4$ |
| This marks the arrival of modern mankind, with the ability to speak and with the associated mental powers. Subsequent evolutionary changes on earth are of its doing. | | | |

each category. As with Table 6-3, the dates indicated for events in this list are given as ranges of possible values, reflecting the fact that they are matters of some debate among scholars, i.e., they are not precisely known and the events themselves, are not well defined. This is particularly true of event 5, which is taken to be the appearance of Australopithecines, and number 7, the first appearance of archaic *H. sapiens*, i.e. nonmodern forms though having many modern attributes. As recent evidence accumulates, it has become clear that event 7 represents a significant and distinctive evolutionary stage.

Following the lead of our logistic/cybernetic theory we construct a semi-logarithmic plot of the events of Tables 6-3 and 6-4, using the midpoints of the ranges shown for each. This is displayed on Figure 6-5. It is immediately seen that the points fall on two parallel lines; the lines are shifted horizontally by one integer of abscissa between numbers 3 and 4, at point "$\alpha$." If this point were deleted from Table 6-4, and therefore from the figure, the remaining nine event points would all fall on a single line. It appears from this that the relation demonstrated between the evolutionary events on this figure is "self-editing" in that it has no place for events that do not conform to its own principle. In view of our earlier admonition that biological criteria may not exclusively determine the Evolutionary Trajectory, this is not unanticipated.

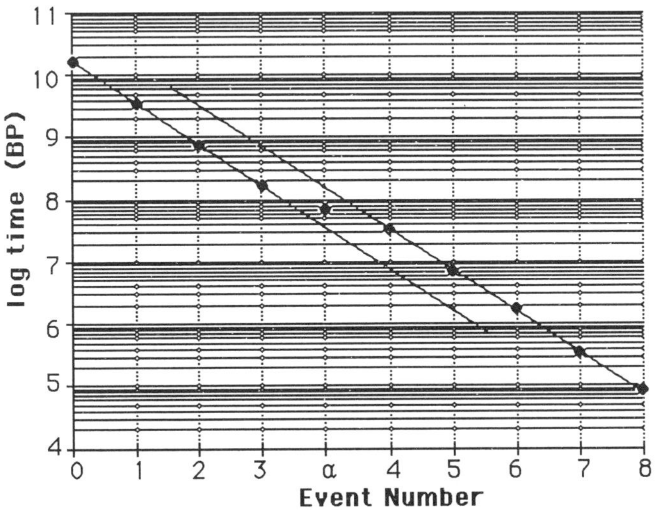


**Figure 6-5** Semi-log plot of natural evolutionary events.

It is particularly noteworthy that point 3, the earliest of the listed phylogenetic classifications, the appearance of mammals, falls next on the line of the events of pre-eucaryotic evolution. Whatever is the internal parameter according to which the events are determined, and that will be discussed later, it has leaped, in a single step, from the eucaryotic radiation to the appearance of mammals. This is indeed troublesome and we may well ask how it can be. For example, almost all paleontologists and biologists would expect the appearance of chordates or vertebrates to be a major transition. And what of the many other important developments of biological evolution, e.g., of the amniote egg that finally freed animal reproduction from its bond to the sea, — are these to be ignored? These omissions, and the range of agreement, will be discussed later when we have established that information is, indeed, the logistic variable of these emergence steps.

TABLE 6-5
EVENTS OF POST-BIOLOGICAL EVOLUTION

| NUMBER | EVENT | DATE (BP) |
|---|---|---|
| 9 | Civilization | $1–2\ (10)^4$ |
| The appearance of purposeful, cooperative behavior in the production of higher forms of organization; a unification of many minds and abilities. | | |
| 10 | Writing | $3–5\ (10)^3$ |
| Rapid and sophisticated communication, begun with event 8, is greatly enhanced, and moved to a higher level of accomplishment by this invention. | | |
| 11 | Printing | $5.45\ (10)^{2}$* |
| Mechanical means for rapidly and inexpensively duplicating the written word, allowing broad dissemination of ideas and information. | | |
| 12 | Digital Communication & Computing | 55* |
| Modern electronics and the computer have already, and promise more, to revolutionize the conduct of our civilization. | | |

* These dates are taken with respect to the calendar year 2000.

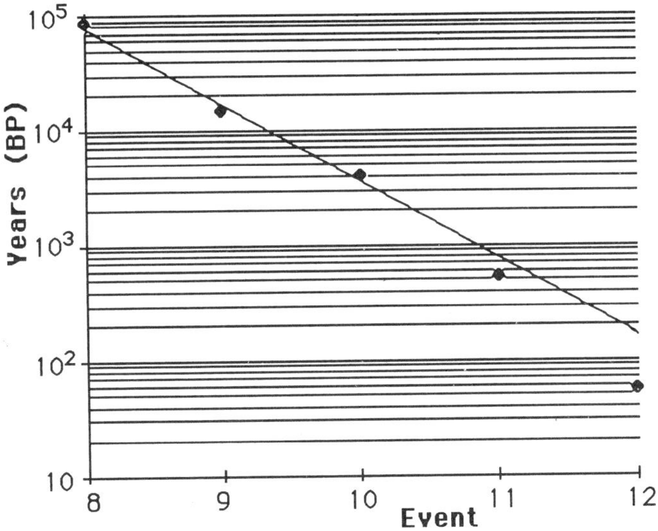


**Figure 6-6** Post-biological evolution.

• POST-ORGANIC EVOLUTION. From the discussion in Chapter 4 we construct Table 6-5, showing the major events of post-organic evolution. As we saw in that discussion, these events are the common choice of scholars of humanity's cultural and intellectual growth.

Following our previous procedure, Figure 6-6 shows a semilogarithmic graph of the events of Table 6-5. For continuity we include event 8, from Table 6-4, and the line on this figure is the extension of the line drawn on Figure 6-5. Events 9 and 10 fall on this line, indicating a continuity of the accomplishments of mankind with the earlier evolutionary events. Here we meet the same problem noted earlier, i.e., the omission of major events that one would ordinarily expect to find, such as the invention of the wheel and axle in Sumer, about 4500–5000 BP, or of fossil fuel power and atomic energy. The hidden variable of The Evolutionary Trajectory that has appeared, and which does not include these events, will be discussed in the following chapters.

## When is The Present

In view of this continuity of the line through events 0 through 10 it is appropriate to determine the equation of this line. In appendix II a "least-squares" regression analysis is performed on events 0 through 10.

This is a standard mathematical technique that results in an equation for the line through a given set of points that is the best fit to their values. The result is

$$\log t = 10.2 - (2/3)n \qquad (6\text{-}4)$$

where t is the time (BP) of event number n.

Note that events 11 and 12 fall below the trajectory line in a systematic way. The dates shown for these points on Table 6-5 are measured from the year 2000 and this raises a question of reference that has been overlooked up to now. When denoting times as "years Before the Present (BP)," exactly when, in this geological context, do we mean by the "Present"? An incorrect choice by, say, several hundred years makes no difference in points 0 through 10 because their orders of magnitude are so much greater. In addition, such a relatively small shift can be disregarded because of the great ranges of uncertainties of those dates. However, the matter becomes important for points 11 and 12 because they are recent and better defined. A backward shift of between 100 and 200 years raises these events to the straight line through the other points. Analysis in the Appendix indicates that the addition of 135 years results in their optimum placement on the line of The Evolutionary Trajectory.

This 135 year adjustment means that our reference for the "Present" should be that much in the future, i.e., approximately the year 2135 of our calendar. The necessity of making this shift has great consequences that will be discussed in Chapter 10. With this change we arrive at the corrected Evolutionary Trajectory, shown as Figure 6-7. The vertical bars on this figure indicate the uncertainty ranges from Tables 6-3 through 6-5.

Whenever we examine a graph, such as Figure 6-7, we should be aware of possible limitations. There may be actual departures of the events from the linear, semilogarithmic occurrence; there are experimental errors, for example, of the date measurements or in categorizing fossil remains; the events themselves are not well defined. A logarithmic representation is rather forgiving of such errors since differences are considerably reduced when their logarithms are compared. For example, the Big Bang is variously estimated to have occurred between the extremes of $1(10)^{10}$ and $2(10)^{10}$ BP. Although these differ by a factor of 2 the logarithms of these dates are, respectively, 10.0 and 10.3, a difference of only 0.3. Probably the largest relative range is that of event 7,

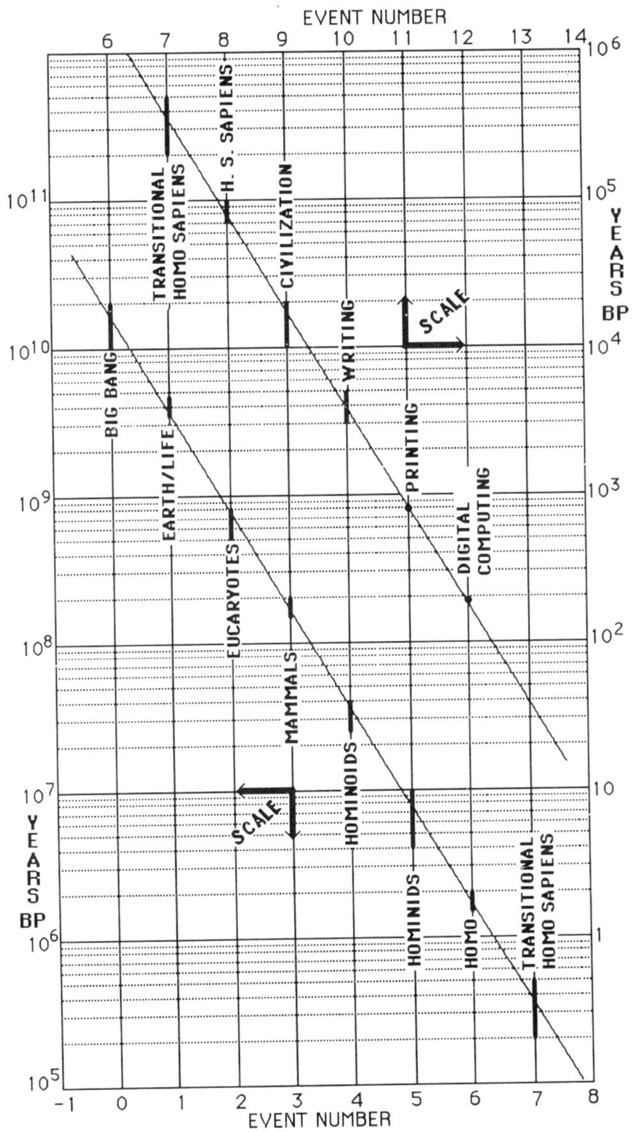


**Figure 6-7** The Evolutionary Trajectory. (Time reference is calender year 2135.)

the appearance of Transitional *Homo sapiens* between $2(10)^5$ and $5(10)^5$ BP, corresponding to a logarithmic difference of 0.4. The logs of these two examples differ from their mean values by only ±0.15 and ±0.2, respectively. Such differences are visible on Figure 6-7 but are not so great as to move the points beyond being considered to lie on the line.

As a result, though the individual points may not fall precisely on the line their deviations are not excessive. However, this can lead to some skepticism that the seemingly good fit is not real. It is important to note that the acceptability of the evolutionary trajectory of Figure 6-7 is not exclusively due to the conformance of its individual data points. Rather, it is based on the collective colinearity of the points, spanning 8 orders of magnitude of time and including 13 events. To a mathematical scientist, the existence of such an extensive relation is *prima facie* evidence that there is a true relation between the coordinates it involves. Furthermore, its existence gives an *a posteriori* rationale for the omission of many other points of organic and post-organic evolution that might otherwise be considered to have equal or greater value according to other criteria. These issues will be discussed more fully in Chapter 8. Of course, this raises the question of just what are the inherent criteria according to which the Trajectory has meaning. In view of the apparent pervasiveness of those criteria and conditions, their identification is vital.

Those recent events that are on the Trajectory illuminate the nature of its hidden evolutionary variable. In the following chapters we consider it to describe a growing information storage and handling capability, first of evolving life forms, and then its extension through the developments of humanity. This will be linked to entropy change during the first, cosmic, development cycle and we will also examine the relation of other measures of change that are considered to be relevant, such as complexity, entropy, and the behavior of natural growth processes. Also to be considered are a quantitative statistical limit imposed by the smallness of numbers on any overall analysis of the evolutionary process, and an imminent limit on that process.

# 7: ENTROPY

## Classical Thermodynamics[1]

Entropy has been called the universal arrow of time. Because of this, a scientifically literate person examining the Evolutionary Trajectory, with its tremendously long time scale and its vast inclusion of phenomena, cannot help being alerted to the possibility of a relationship involving entropy. It therefore behooves us to explore the nature of this unique quantity. In this chapter and the next we will develop the relationship between entropy, information, and that trajectory. We begin with a background examination of thermodynamics, the study of equilibrium states of matter, particularly when they undergo exchanges of energy. As a science, it can be dated back at least to 1798 when Count Rumford, while boring cannon barrels for the Napoleonic wars, measured the rate at which mechanical work is converted to heat. Since then a great deal of fundamental and applied, and behavioral and theoretical information has been gathered and interpreted, and a considerable "handbag" of experimental and mathematical rules and tools has been developed. To be taken seriously, any scientific proposal made today must satisfy the strictures imposed by thermodynamics. One of the most important discoveries of thermodynamics is the existence of entropy.

Physical systems are described by a number of variables. A mass of gas, such as the air about us, has a pressure, P, expressed in units of $lb/in^2$ or $Newtons/m^2$, it has a volume, V, in liters or $m^3$, and it has an absolute temperature, T, in Kelvin. These properties are generally interrelated so that changing one of them results in changes of the others. Chemists have long described the phenomenological behavior of gasses with the empirical Laws of Boyle, and of Charles, and of Gay-Lussac. These can be combined into the statement that for a fixed amount of gas:

$$PV/T = \text{constant.} \tag{7-1}$$

This is called the Equation of State of an Ideal Gas. Although it is usually a good description of real gas behavior, it is not accurate under some conditions, such as very low temperatures or very high pressures. We see from Equation 7-1 that if the temperature is held constant then the product of pressure times volume is also constant and, for the fixed amount of gas, the pressure must increase as the volume decreases. This is just what we find when we collapse an empty, i.e. air filled, tightly sealed container. Its temperature is fairly constant because the container is at room temperature. The container volume decreases as we squeeze it so its air pressure increases. This continues until the pressure becomes great enough that it either blows off the cover or ruptures the container or it opposes our applied force so we can't collapse it any more.

In the study of thermodynamics, the quantities used to describe matter occur in what are called conjugate pairs. These have the property that their product is energy. When we collapsed the container we did work on it, and work is energy in transit, here going from us to the air/gas. The work we do when we change the volume by a small amount $\Delta V$ is equal to the product $P \times \Delta V$. (Recall that the Greek letter "delta," $\Delta$, denotes a change of the quantity that follows it.) The total work is the sum of all these small changes, with P increasing slowly with each consecutive decrease of V. The conjugate pair {PV} represents mechanical work. Among other conjugate pairs known to thermodynamics are {HM} — the magnetic field strength and magnetization of magnetic bodies — for magnetic energy, and {Vq} — electric potential and charge — for electric energy. Physicists and engineers of the nineteenth century were bothered by the fact that, while heat is a very general form of energy that affects all systems, there were no conjugate variables to describe it directly. Since heat flows from hot to cold bodies, the temperature must be involved, but there seemed to be no candidate with which temperature was conjugal.

During theoretical studies of thermodynamic systems, it was found that a small amount of heat, $\Delta Q$, changes the system description as much as if the volume or pressure were altered. The change depends on $\Delta Q/T$ and it appears that the quotient $\Delta Q/T$ behaves like a physical property of the system, equivalent to P, V, H, M, etc. If we denote this new variable by S then the small change of S is $\Delta S = \Delta Q/T$. Therefore $T \times \Delta S$ is $\Delta Q$, the heat energy. S is called entropy from Greek for "change of energy." Entropy, then, is the conjugate of temperature since their product is heat energy.

Entropy is a descriptive variable of thermodynamic systems and when its changes were studied it was found to possess a distinctive characteristic. Consider an isolated body or system, that is, one that cannot interact with its surroundings, e.g., it cannot exert a force or exchange energy in any form. It was found that the entropy of such a system will always be the maximum value it can achieve, consistent with any restrictions on the other state variables: P, V, T, H, etc. Any property that attains a maximum, or minimum value, or remains constant, under a very wide range of conditions, is extremely useful and important in analyzing and understanding behavior. For example, mechanical interactions are guided by the conservation of momentum — unless external forces are applied, momentum remains constant through complex mechanical changes. This is the basis of one of Newton's laws of mechanics, that the rate of change of momentum of a mechanical system is equal to the force applied to it; without the force, its momentum remains constant. Similarly, the stability of a material system can be understood and analyzed by considering its available, or free, energy, which always attains a minimum value.

The existence of entropy as a function of the state of a system, and the fact that it attains a maximum value, are statements of the two parts of the Second Law of Thermodynamics. (The first law states that there is another function, the internal or free energy, that is also a function of the state variables.)

Besides establishing the maximal property of entropy in noninteracting systems, thermodynamicists also studied the changes that occur when systems do interact, taking account of all changes of all the interacting parts and of their surroundings. Reasoning from very basic considerations, they have shown that if an amount of heat energy, $\Delta Q$, is exchanged the net entropy always increases by $\Delta S = \Delta Q/T$. Since this increase occurs for all interactions it follows that the *total entropy of the universe is increasing*. Thermodynamics was therefore provided with a valuable new tool that changes in only one direction as time advances. Strange as it may seem, and contrary to our common experience, there is no other physical property that has this behavior. We are certainly aware that ontogeny, the process of fetal development, is unidirectional, as is the cumulative progress of evolution itself. If we are to relate these to fundamental, underlying, physical processes we therefore expect that relation to be through a similarly directional physical mechanism, and that connection must, therefore, involve entropy.

To pursue this linkage we investigate the fundamental nature of entropy, for we note that, as valuable as was the discovery of the ever-increasing behavior of entropy, this only gave a behavioral description. For a deeper meaning we must turn to the subject of statistical mechanics.

## Statistical Mechanics[1]

At the same time as the developments described in the previous section, research into the properties of matter was also proceeding along other lines. In particular, theoretical physicists of the nineteenth century were challenged by the problem of understanding the detailed behavior of the ideal gas. This gas was imagined to consist of atoms that are very small, hard spheres, moving rapidly and randomly throughout their enclosing volume, colliding with each other and with the container walls. Theoretical techniques of mechanical analysis were quite advanced and very sophisticated, and one could imagine writing equations describing the motions of all the atoms. If the atomic positions and velocities were known at any instant, it should have been possible to calculate the entire future kinetics. From this the pressure of the gas can be determined because P is the force exerted by the particles due to collisions with an area of the wall. The absolute temperature can also be found since T is proportional to the average kinetic energy of motion of the particles. The gas properties would be completely known.

The difficulty in carrying out these calculations was not with predicting the future motions but with keeping track of the $10^{24}$ atoms in a liter of gas. Even the largest and most rapid computers today aren't up to this task, and we are talking about 100 years before the computer. The investigators were therefore forced to fall back on statistical means, dealing with average properties and the deviations from those averages. Their accomplishments were, nonetheless, striking. They analyzed the theoretical ideal gas and arrived at the result that PV/T = constant, the same relation that was already known to describe the behavior of real, low density gasses.

The statistical approach was carried further by studying departures from the average behavior. We illustrate this by considering the distribution of gas atoms in the volume it occupies. To simplify our discussion let us divide the volume into three distinguishable regions: I, II, III, along

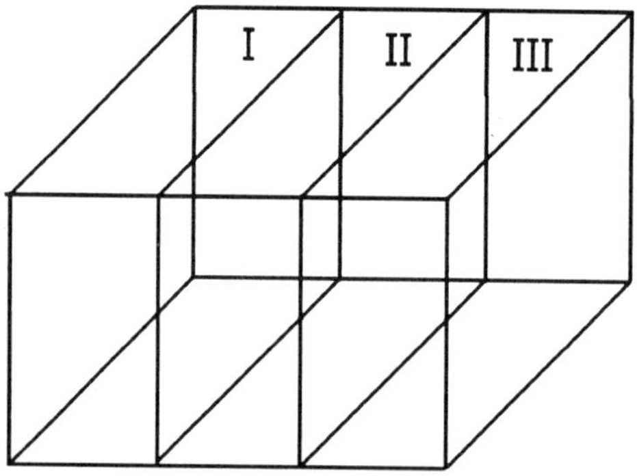


**Figure 7-1** The volume of a gas, divided into three equal parts.

one direction, as shown in Figure 7-1. Furthermore we take our hypothetical gas to consist of only three atoms: a, b, c. All the possible arrangements of three particles in the three regions are shown in Table 7-1. Each of these is called a microstate. Thus, arrangement 2 has atoms, a and b, in volume I while atom c is in volume II. This is distinct from arrangement 3, in which atom c has moved to region III, and from arrangement 4 that has interchanged atoms b and c from arrangement 2.

When we examine a real gas we cannot differentiate the atoms so that the three particles a, b, c are indistinguishable. It is therefore preferable to

TABLE 7-1
DISTRIBUTIONS IN REAL SPACE REGIONS

| ARRANGEMENT (microstate) | REGION I | REGION II | REGION III |
|---|---|---|---|
| 1 | a, b, c | – | – |
| 2 | a, b | c | – |
| 3 | a, b | – | c |
| 4 | a, c | b | – |
| 5 | a, c | – | b |
| 6 | b, c | a | – |
| 7 | b, c | – | a |
| 8 | – | a, b, c | – |
| 9 | c | a, b | – |
| 10 | – | a, b | c |
| 11 | b | a, c | – |
| 12 | – | a, c | b |
| 13 | a | b, c | – |
| 14 | – | b, c | a |
| 15 | – | – | a, b, c |
| 16 | c | – | a, b |
| 17 | – | c | a, b |
| 18 | b | – | a, c |
| 19 | – | b | a, c |
| 20 | a | – | b, c |
| 21 | – | a | b, c |
| 22 | a | b | c |
| 23 | a | c | b |
| 24 | b | a | c |
| 25 | b | c | a |
| 26 | c | a | b |
| 27 | c | b | a |

consider states of the system to be determined by the numbers of particles in a region, rather than exactly which ones they are. As a result the configurations having two different particles, e.g., (a,b) or (a,c), in region I are now equivalent and are considered to be different manifestations of the same "macrostate." If we denote all the particles by circles, rather than by letter identification, we can count the number of individual arrangements — the number of microstates — corresponding to each macrostate. This is shown on Table 7-2. On that table the subheading of the column giving the number of arrangements is the technical term "degeneracy," meaning a lack of variability or of variety, it denotes the number of microstates that are equivalent in constituting a particular macrostate.

As all three regions in Figure 7-1 are the same size we assume that there is no reason for the particles to prefer one region or another. Therefore, all the microstate arrangements of Table 7-1 are equally likely to occur, so the probability of each macrostate in Table 7-2 is simply equal to the fractional number of microstate cases it includes. The probability of a macrostate equals its degeneracy divided by the total number of cases. For example, the probability of finding the system in macrostate 10 is $P_{10} = 6/27 = 0.2222$ and of state 6 is $P_6 = 3/27 = 0.1111$; we see that state 10 has the highest probability. This state has an equal number of particles in each of the three volumes of space, which is the closest we can come, with a three particle-three region system, to having a uniform distribution. This is an example of the general result that a uniform distribution of particles is the most likely state of a system.

TABLE 7-2
DISTRIBUTIONS IN REAL SPACE REGIONS

| MACROSTATE | REGION I | REGION II | REGION III | NO. OF CASES (degeneracy) |
|---|---|---|---|---|
| 1 | ooo | – | – | 1 |
| 2 | – | ooo | – | 1 |
| 3 | – | – | ooo | 1 |
| 4 | oo | o | – | 3 |
| 5 | oo | – | o | 3 |
| 6 | o | oo | – | 3 |
| 7 | – | oo | o | 3 |
| 8 | o | – | oo | 3 |
| 9 | – | o | oo | 3 |
| 10 | o | o | o | 6 |

Note that in this discussion we have divided the volume and considered the distribution in only one direction. The same results must be true in any direction so that the particle distribution with the greatest probability is completely uniform throughout the volume.

Of course, the existence of non-zero probabilities for the other states means that the particle motions do sometimes take them into non-uniform states. However it is a very general statistical result, known as the Mean Value Theorem, that as the number of particles in a system becomes large, the probability of the uniform state increases greatly, relative to the other cases. For a real gas with a very great number ($N \sim 10^{24}$), the particle distribution can always be considered to be uniform; there is no clumping of molecules and no regional preference.

## The Arrow of Time

The essential connection of these statistical considerations to thermodynamics was made by Ludwig Boltzmann (1844 to 1906). He postulated a relation between the degeneracy, W, of a state of a system, and the entropy, S, of that state. We can write that relation as

$$S = 2.30259\ k \log W = k' \log W \tag{7-2}$$

where $k = 1.38032\ (10)^{-23}$ joule/degree is called Boltzmann's universal constant.[A] Since the degeneracy of a macrostate is proportional to the probability of finding the system in that state we can accept the conclusion that Equation 7-2 may also be expressed in terms of the probabilities of its various microstates. This is written in the form

$$S = -k' \Sigma\ P_i \log P_i \tag{7-3}$$

where the Greek letter sigma, $\Sigma$, denotes a summation of all the microstates of the macrostate, each denoted by its own value of i. Thus, each of the three microstates that constitute macrostate 8, on Table 7-2, are equally likely and therefore have $P_i = 1/3$. Equation 7-3 then gives: $S_8 = -k' \Sigma(1/3) \log(1/3)$. The sum has three terms so that $S_8 = -k' \times 3 \times (1/3) \times \log(1/3) = k' \log 3 = 0.477\ k'$. For macrostate 10 we find $S_{10} = k' \log 6 = 0.778\ k'$. Therefore, the entropy of state 10 is 0.301 k′ greater than that of state 8.

Real physical systems have extremely large numbers of components. As a result their degeneracies and entropy differences are much greater, making it (almost) impossible for the system to be in any state other than its most probable or most degenerate one, that is, in the one with the highest entropy. Furthermore, if the system changes it will seek a new condition of maximum W and S, and if these cannot increase they will remain unchanged. This means that S does not decrease.

As time goes on all things do change. Locally and on a large scale all natural systems are varying, and undergoing alteration, and exchanging energy. This is true for the gasses we've used as an example; it holds for living things, that grow, age, and decline; it applies to land masses, that erode, spew forth volcanic material, and quake; and it is so for stars and galaxies, that burn atomic fuel, radiate, explode, and decay. In all of these changes the entropy will increase so that the entropy of the entire universe is increasing. We have seen that the highest entropy state has a uniform distribution of particles in space. This means that the universe must be moving toward a condition of complete uniformity in the distribution of matter. The molecules of the tree outside your window, are now localized in its trunk and leaves. Given enough time, they would eventually be spread uniformly throughout the cosmos, along with those of the moons of Jupiter, those of the distant galaxies, and those of you and of me. The increase of entropy implies a movement toward uniformity, toward random motion of fundamental constituents, toward chaos. This ultimate winding down through thermodynamic decay, or entropy growth, is often referred to as the entropic, or thermodynamic, or heat "death" of the universe.

Yet, as we look around it seems that not everything is winding down. That tree outside your window grew and proliferated, ordering its chemical constituents into cells, and differentiating its parts into bark, leaves, and flowers. The same is true of the bird in the tree and of you and me, who are looking at them and appreciating their beauty. It would appear that life, in its various forms, has managed to reverse the law of entropy. How can this be? The answer is that the law still applies, but in a larger sense. To illustrate this let us consider the manufacture of some common object, such as an automobile or a toaster or a brick. For these we must have taken mineral components that are dispersed through the earth, and concentrated and refined and processed them into an extremely ordered arrangement that functions as a system. We have undone nature's randomizing. The probability that these minerals will naturally be in the form of a toaster is very much smaller than their being randomly dispersed in the earth's crust. We have lowered their degeneracy and entropy.

A closer look, however, shows that we can not make the toater without doing many other things. Moving the minerals, refining and foming them to shape, etc., require the expenditure of a considerable amount of energy from motors and engines. This energy is derived from fuels of various sorts and extracting it involves, ultimately, a flow of heat. Since the basic, thermodynamic definition of entropy is heat exchange: $\Delta S = \Delta Q/T$, all these processes increase entropy. Surely we lowered the entropy of some set of components when we organized them into the toaster. However we increased the entropy of the surroundings by an even greater amount. We have created local order at the expense of greater disorder.

The same considerations apply to living systems on the earth. They lower their own degeneracy by forming into statistically unlikely configurations. For this they use available local energy sources that have, primarily, derived their own energies from the sun. Such a system with its interacting surroundings is called the "local universe." In supplying this energy the local sources, including the sun, are moving toward greater disorder. The overall local universe is increasing entropy at a much greater rate than the living systems are lowering it.

Prigogine[2] has pointed out the general property that both order and disorder result from entropy generation. A particular system that interacts with its surroundings has no restriction to increase its own disorder and raise its own entropy, although the entropy of the entire local universe must increase. The creation of local order, and local entropy reduction, is a central feature of the evolutionary process and we will return to it in the next section and in subsequent chapters.

Even if we ignore the ultimate heat death of the universe, the overall process of entropy increase is important. It is a property that changes in only one direction as time advances. We have already commented that, contrary to our senses and experience, there is no other dynamical law of physics that has this property. The others would apply equally well if time were to run in reverse. Consider an example involving Newton's Law, the fundamental law of mechanics. Imagine a movie film showing a ball being tossed into the air. It rises and slows, eventually reaching a maximum height and then falls into the hand of the catcher. If the film is shown in reverse we see exactly the same sequence and can't tell which is forward and which is backward. On a microscopic, atomic scale this is true of the laws of mechanical motion that govern how all things change.

Now examine a slightly more complicated situation. Imagine that there are two, touching billiard balls on a pool table. The white cue ball approaches and strikes them, causing them to move apart. One falls into a side pocket of the table and one rolls out of view while the cue ball, having imparted its momentum to the numbered balls, remains stationary at the point of impact. Such billiard shots require some skill to accomplish but are fairly common. If the sequence is reversed we see one numbered ball leap out of the pocket while the other rolls into the field of view. The two come together, striking the stationary cue ball, which then rolls away while the original two balls become stationary and in contact. There is nothing in this second sequence that violates a law of behavior. Just as we did not seek the cause of cue ball motion in the first case so we should not question the motion sources in the second. Yet, our experience tells us that the second case is unlikely. Its probability is small enough that we know it to be a reversed time sequence. This reflects one sense of meaning of the increase of entropy. Systems move to their most probable state, thereby increasing their degeneracy and entropy. Opposite changes are allowed but are unlikely, i.e., they have lower probabilities. Particularly in large systems, the lower probabilities of the reversed individual events multiply together to produce a vastly unlikely sequence. Such an occurrence is easily recognized as being false or contrived.

Unfortunately we are limited creatures and our senses do not allow us to recognize entropy inherently. Imagine we are watching a fire with a friend. The burning log emits light and our experience tells us that objects which radiate spontaneously are generally hot. They can cause burns and are better left at some distance. As the fire dies the embers darken and, we know, cool until we feel safe in approaching more closely. Now our friend is from a distant galaxy and has no knowledge of temperature. However, he can detect entropy. His senses indicate that the burning log has great chaotic activity of its outer surface and is imparting it to the air. It has a high level of entropy generation. He knows that such entropy sources are destructive and dangerous and are better left at some distance. As the fire dies the entropy production level decreases until he feels safe in approaching more closely.

The actions of our alien friend were the same as ours. He can sense entropy as well as we can sense or measure heat. To him, as to the thermodynamicist, entropy is a real property of a system, as much as its temperature, pressure and volume. The thermodynamicist is at a disadvantage compared to the alien because he cannot judge entropy directly. Just as he measures temperature by the change in length of a thermometer column, or some equivalent means, he calculates entropy from changes of other material properties. That the total entropy of events does not decrease allows him to assign a direction to the flow of time. It is in this sense that entropy is referred to as an indicator of "the arrow of time."

## The Entropy of Living Systems

We have already indicated that the Second Law of Thermodynamics does not prevent the local decrease of entropy that occurs when living systems form ordered structures. Beyond that we now discuss the fact that it is *required* to occur. It is only in the past few years that some understanding had developed of how living systems might be able to reduce their entropies, seemingly in opposition to the general dictum of the Second Law. To begin we must recognize that these systems are in non-equilibrium states. A gas, such as we considered earlier, is either completely isolated from or in equilibrium with its surroundings. In both cases there are no influences trying to chnage it; expect for fluctuations, such as we saw for the 3 particle gas in a 3 section container and which are insignificant in a real gas, it is unchanging.

By contrast living things are not isolated; they exchange energy and matter with their surroundings (because of this they are called "open" systems), allowing them to perform metabolic functions. And they are not in equilibrium, for example they may maintain a different temperature from their environment. There are forces imposed on them to change and revert to a nondistinctive (dead-?) state and it is the process of metabolism that allows life to resist these forces and to maintain its nonequilibrium status. The internal processes develop structures, mechanisms, and opposing forces that maintain the lifeform in its dynamic state of balance. Rather than equilibrium with the environment we have a dynamic, nonequilibrium, steady-state.

In this steady state, energy absorbed from the environment is used to drive the metabolic process. Thus, sunlight consists of photons of the proper energies to activate electron transitions that, through photosynthesis, generate chemical actions of use in maintaining plant separateness. Ultimately, after it is used, the energy of the chemical reactions is surrendered by the plant, through reradiation or conduction to the air or ground, in a form that is degraded and not usable.

Energy transfers are equivalent to heat transfers and since an entropy change, $\Delta S$, occurs when heat, $Q$, flows at temperature $T$: $\Delta S = Q/T$, they can be said to be accompanied by an entropy flow. The energy absorbed by plants approximately corresponds to the temperature of the sun (about 5000K) so we have $\Delta S_{incident} = Q/T_{sun} = Q/5000$. When the same heat is emitted it is at the temperature of the living body (approximately 300K) so that $\Delta S_{out} = Q/300$ is much larger. The increase of entropy is approximately $Q/300–Q/5000 = 0.00313Q$. This is consistent with the anticipated, universal entropy increase.

There is an additional entropy to be considered because the living system uses the chemical changes induced by the energy it is processing to support and enhance its ordered state. The highly structured condition of a lifeform is far from equilibrium and therefore has greatly reduced degeneracy, greatly lowered probability, and very small internal entropy compared to the equilibrium condition of a hypothetical, random distribution of its constituents. Compared to this uniform, nonliving distribution the living system can be said to have a negative entropy, $-\Delta S_{internal}$. This relation is often expressed by stating that life has the ability to lower its entropy or, alternatively, to extract "negentropy" (negative entropy) from the environment of energy processes.

The mechanism by which the $-\Delta S_{internal}$ is generated, allowing life forms to exist in orderly, structured configurations, has long been the subject of attempts to reinterpret aspects of the Second Law. That law deals with equilibrium states of a system so these reformulations have emphasized the response of systems that are being driven away from equilibrium. In particular, we consider the forms in which Schneider and Kay,[4] following Swenson,[5] have recently restated the Second Law:

> As systems are moved away from equilibrium, they will utilize all avenues available to counter the applied [forces]. As the applied [forces] increase, so does the systems ability to oppose further movement from equilibrium.

In this quotation we have substituted the word "forces" for the original, more technical word "gradients" meaning changes of magnitude. For example, a thermal gradient, i.e., a rise of temperature between one body and a warmer one, imposes an effective force causing heat energy to flow between them; a gradient of electric potential (voltage) generates an electric force that causes charge to flow as electric current; a pressure gradient in a fluid causes mass flow. In particular, there is a gradient of energy between the sun and earth, with earthly systems, living and not, acting as the transduction means for its degradation to a lower form.

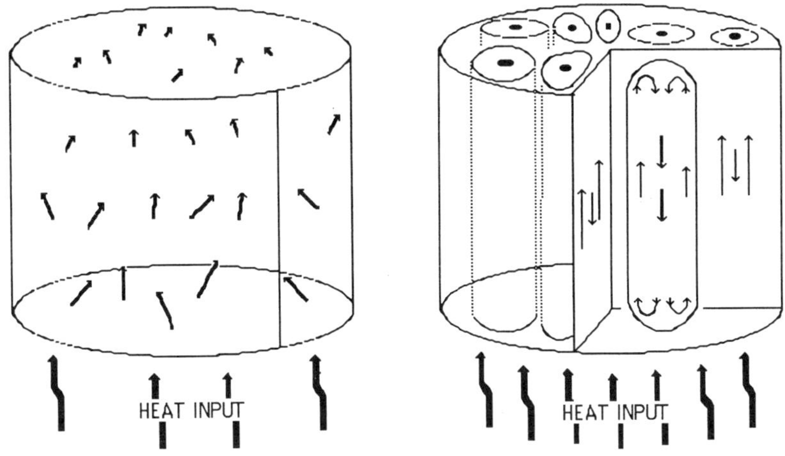


**Figure 7-2** Benard Cell structure. (a) With low heat input the flow is by conduction. Internal arrows denote the decreasing molecular drift velocities from bottom to top. (b) With greater input an ordered structure is established by the creation of convection cells.

This restatement by Schneider and Kay postulates that the harder a system is driven by such gradients and their forces, the more it changes to oppose the forced changes. To illustrate this let us consider a Benard cell; this is a container of viscous fluid that is heated from below, as shown in Figure 7-2a. Initially the fluid is uniform and in equilibrium with its surroundings, with its molecules moving randomly. Over a low flame the energies of those molecules near the bottom will increase. Since their new energy is kinetic they move more rapidly and transfer some of it to other molecules with which they collide. The temperature gradient between bottom and top, through the mechanism of this molecular motion, causes heat energy to flow upward and eventually to be carried away by the air above the cell. This is a "linear" heat flow mechanism, meaning that the increase of kinetic energy and the temperature gradient are directly proportional to the rate of heat input.

The mean distance between molecular collisions limit the distance and rate of heat transfer by this linear mechanism. Consequently, if the heat input is increased beyond this limit, nonlinear mechanisms must manifest themselves. In this case small groups of molecules can gain heat together and move through the liquid. Because this hot fluid is less dense than its surroundings it rises to the surface as an entity; it is less likely to dissipate its energy through intermediate collisions and can deliver it more rapidly to the air above the container. As the input increases more a second group is formed at some point on the base as soon as a previous one has moved away, and then a third, and so forth. Suddenly a transition of behavior occurs; it might be said that the system has passed through a bifurcation point. The result is a column of warm water rising to the surface, curling over as it surrenders its heat to the air above, and then, cooled, descending. The locations of these initial hot-spots is random from the macroscopic perspective but different points of the container arrive at this transition simultaneously and several circulating columns are formed. In this way the heat transfer has changed from conduction to convection.

Note that the cell now contains an ordered arrangement of circulating fluid columns, as shown schematically in Figure 7-2b, each proceeding more or less independently. It has organized itself to reduce the gradient that drove it from equilibrium. This state has a lower *a priori* probability of occurring than the random, homogeneous motion of the molecules. The system has thereby reduced its internal entropy. Such self-organization plays a central role in the theory of complex, hierarchical systems. In Chapter 5 we discussed the system transitions that appear as emergent events of an evolutionary process. Those transitions were described in terms of the system passing through a bifurcation point. The discussion of the Benard cell indicates, by analogy, how such changes occur through changes of the thermodynamic, non-equilibrium, steady-state.

Ultimately, if the heating is increased further, even the rising columns become inadequate to convey the heat. Their temperatures rise until fluctuations cause superheated groups of molecules to form internally. These are bubbles of vapor that rise rapidly to the surface; in this way the cell passes through another bifurcation point and boiling begins.

Note that, in this case, we have been able to describe the behavioral bifurcations in sufficient detail that they can each be understood as a process of emergence to a new structure and behavior. It has been suggested that this is analogous to what occurred early in the history of the earth.

> The free energy gradients (solar and geothermal) of prebiotic Earth provided a structuring environment for the generation both of thermodynamic potential and molecular complexity. Energy gradients established an initial condition of nonequilibrium. Under that condition, putting small entities (biomonomers) together to make larger ones (biopolymers) produced thermal entropy

> absorbed by the sink of space. The aggregation of proteinoids into microspheres under the hydrophobic tension of an aqueous environment also created entropy at the expense of free energy. These microspheres provided a selective context for the honing of the funtional relationships that would lead, under selection, to living systems.[6]

The details are, of course, unknown but the thermodynamic concept is surely appealing. Once this initial step had been taken the primitive structures that resulted contained means to store their form and replicate it. As a result they were not so easily destroyed as the Benard cell structures. They then constituted a new element in the ecosystem and over eons they altered it. This, in turn, produced new gradients to which they were subject and these induced further changes, i.e., the living systems evolved.

This discussion suggests that, at least in its early stages, the evolutionary transitions involved local reductions of entropy. In the next chapter we show that such reductions are equivalent to increases of information, where that term has a specific technical meaning that is only slightly different from its generally understood meaning. The evolutionary process is therefore made equivalent to the growth of information.

## Endnotes

[A] The usual expression of equation 7-2 is: S = k *ln* W where *ln* denotes "Naperian" of "natural" logarithm, i.e., to the base e = 2.71828 rather than the base 10. To use "log" instead, it is only necessary[3] to replace k, by k′ = 2.30259 k. S = k *ln* W is carved on Boltzmann's tombstone, memorializing the great significance of his insight.

# 8: INFORMATION

## Definition

Modern communication is largely based on binary logic, in which messages are transmitted using only two discrete signal levels. In the binary number system these levels denote 0 or 1 and any number can be expressed as a series of 0's and 1's if we understand them to represent decreasing powers of 2. For example, we have $26 = 16 + 8 + 0 + 2 + 0$; term by term this is identical to $26 = 1 \times 2^4 + 1 \times 2^3 + 0 \times 2^2 + 1 \times 2^1 + 0 \times 2^0$. In binary, digital code this yields a number, 11010 derived from the coefficients of the powers of 2. It is said to have a length of five binary units, or "bits." We should note that, except for the use of the base 2 instead of the base 10 this is exactly how we commonly write numbers in decimal units. Thus, $3072 = 3 \times 10^3 + 0 \times 10^2 + 7 \times 10^1 + 2 \times 10^0$.

Communications engineers are concerned with designing efficient ways to transmit messages. To optimize their efforts, they employ a quantity called "informational entropy" to define the information content of a message. Binary units are also used in expressing this entropy[1]

$$S = -\Sigma\, P_i \log_2 P_i \quad \text{bits} \tag{8-1}$$

where it is understood that the logarithm to the base 2: $\log_2 P_i = y$ means that $P_i = 2^y$, just as we previously had $\log_{10} P_i = w$, meaning that $P_i = 10^w$.

Let us imagine that an information engineer is measuring some unknown quantity and knows only that it ranges between 0 and 7. The possible three bit numbers that cover this range are:

| binary | | digital |
|---|---|---|
| 000 | ⇔ | 0 |
| 001 | ⇔ | 1 |
| 010 | ⇔ | 2 |
| 011 | ⇔ | 3 |
| 100 | ⇔ | 4 |
| 101 | ⇔ | 5 |
| 110 | ⇔ | 6 |
| 111 | ⇔ | 7 |

Because he doesn't know the correct value he must assume that each bit can be either 0 or 1 with equal probability of 1/2. The probability of any three place number is $P_i = 1/2 \times 1/2 \times 1/2 = (1/2)^3 = 1/8$. We then have $\log_2 P_i = \log_2(1/2)^3 = 3 \log_2(1/2) = -3$. [Here we have used the relation $\log_2(1/2) = \log_2(2^{-1}) = -\log_2 2 = -1$.] As the sum in Eq. 8-1 includes all 8 possible answers, it gives $S = -\Sigma(1/8)(-3) = -8 \times (1/8) \times (-3) = 3$ bits. We recall that such a uniform distribution, i.e., with uniform probability for all outcomes, is the most random situation, and has the highest entropy.

The engineer then finds the actual value of his measured quantity to be, say, 011 ⇔ 3. After the measurement he therefore knows that $P_3 = 1$ and $P_i = 0$ for all other subscript values. Eq. 8-1 then gives $S = 0$. The entropy, or the degree of uncertainty of the answer, has decreased from 3 bits to zero and the engineer can state that through the measurement he has gained 3 bits of information.

This is a special case of having n equally likely binary outcomes (i.e., each can be yes or no, 0 or 1, + or –, etc.). The initial entropy is n bits and, once a specific value is found or a message is received specifying that precise value, the final entropy is 0. The information gained by determining that final value is therefore n bits. In engineering communications books, Equation 8-1 is referred to as the information content of the message. Even if the message is not in binary form, but all outcomes are still equally likely, this equation still gives the information content to be $-\log_2 P_i$ bits.

In the discussion of the previous chapter we were able to distinguish the forward from the reverse pocket billiards sequences by inference from the probabilities of their occurrence. Statistical likelihoods provided us with critical information to distinguish the forward direction of the time sequences. The same would be true for any physical change or interaction. As a result of the relation between the thermodynamic entropy and probability, expressed by Equation 7-3:

$$S = -k' \Sigma P_i \log P_i \quad \text{(joules/degree)}, \tag{8-2}$$

we can understand why entropy imparts the same information. The summation, $\Sigma$, in this equation, and the methods of the last chapter, require enumeration of all the individual events, configurations and probabilities if we are to calculate S. Unfortunately these are not known in most real cases. However, the direction of increasing entropy can frequently be calculated from the alternative definition of its relation to the system's Thermodynamic properties. As a result, for reasons of simplicity of calculation and clarity of physical meaning, the entropy is generally preferred over the raw probabilities as an indicator of time flow. It is frequently regarded as being a more fundamental quantity than the microstate probabilities.

We immediately note the similarity of Equations 8-1 and 8-2, even though one involves logarithms to the base 2, the other to the base 10, and one has a unity coefficient while the other has k′. Both of these differences are merely equivalent to a single multiplying factor and this is reflected in the fact that they are measured in different units. However, because of this, and their different uses and domains of application, it is often believed that the thermodynamic and informational entropies are different, incommensurate quantities. However physical and mathematical scientists recognize that two quantities whose functional-mathematical dependencies are the same but which differ by only a numerical constant are really a single quantity being measured in different units.[2] The fact that such a quantity may have different origins, interpretations, and conditions is an acknowledgement of the robustness of the concept it represents and of the power of mathematics to recognize the underlying commonalty.

In this particular case, specific quantitative analysis[3] by physicists has established that entropy and uncertainty are the same and that information is the reduction of uncertainty, making entropy and information complements of each other. Yet, despite this, there remains a mystique connected with the applications of these concepts to life. As discussed earlier, the steps in the origin of life are still very uncertain, with several scenarios being extant. Even in very general principle the process is poorly understood. This, plus the clear difference between the inanimate and animate worlds, makes it common to assume that the natural laws governing these two states are different or, at least, have different actions. The old, rejected, idea of an *elan vitale*, a special property of life, while not acknowledged as such, is implicit in several studies of entropy and information in living systems. D. R. Brooks and E. O. Wiley[4] have suggested that only organisms have "Instructional information" that is transcribed and transmitted between generations and through which an organism builds itself, and that the changes and restraints on this form of biological entropy are "of fundamental importance to understanding the basic process of evolution." Similarly, R. U. Ayres[5] suggest that information has a component that he calls "Survival Relevant," consisting of "Survival Useful" and "Survival Harmful" parts; supposedly the progress of organic evolution can be correlated with these.

Just as it is badly stressing the concept of "complexity" to apply it over the entire informational trajectory, so it is different to see how the concepts of internal instructional information, or survival information, or any other similar, special construct, can be extended to the first and the latter logistic cycles of that Evolutionary Trajectory. Furthermore, such special assumptions seem unnecessary.

## Information and Language

The implications of Equation 8-1 go beyond the simple examples given above. As we are discussing information, let us study it by turning to language, the original and most common means of imparting information. If we ignore punctuation and capitalization the English language has 26 letters and a space for separating words, for a total of 27 symbols. If all were equally likely to occur then each printed symbol would have an information content of $\log_2 27 = 4.76$ bits. We realize that this is not an accurate estimate of the actual case because all letters are not equally likely to occur so that their $P_i$ are not all the same. In English every word must have a vowel (or the letter y) and these generally occur with greater frequency than other letters, as do the spaces between words. The most common symbols, in decreasing order, are: e, space, t, o, a, n, s, h, i, r, d, l, u, ... While the $P_i$ of some symbols are greater than $1/27 = 0.037$, e.g., P(e) and P(space) $\approx 0.1$ each, the $P_i$ of others are lower, e.g., P(b) $\approx 0.01$. This results in a lowered maximum entropy from Equation 8-1 and, consequently a lowered information content for any message; for English the resulting average entropy value is approximately 4.0 bits per letter.

The unequal occurrence of letters is only one aspect of the departure from uniformity. Equation 8-1 assumes that all the letters are independent, that is, there is no correlation between their occurrence, although they are not necessarily equally probable. However, as a result of the rules of spelling, certain combinations occur with greater frequency than others. For example, if the letters were independent we would expect the combinations "eb" and "be" to be equally likely, with a probability $P(e) \times P(b) = 0.1 \times 0.01 = 0.001$. Instead, because "be" is a common word and a common root, they are close to P(eb) $\approx$ .0001 and P(be) $\approx 0.03$. Similarly, $p(the) > p(t)p(h)p(e)$.

Due to such higher order couplings, one can, in principle, calculate a nearest neighbor correlation entropy.

$$\Delta S^{(1)} = -\Sigma\, p_{ij} \log_2 p_{ij} - \Sigma\, p_{ijk} \log_2 p_{ijk} - \dots \qquad (8\text{-}3)$$

These terms result from the fact that a random jumble of letters is meaningless so that letters must be constituted into words; words have meaning only when their spellings are standardized. $\Delta S^{(1)}$ constitutes information that resides in the language structure itself.

All this is on the level of the individual letters and ignores the fact that there are even higher levels of coordination in language. These are related in the rules of spelling and grammar, arising because standardized constructions are necessary to interpret what is read or stated. As examples, most sentences have a subject, verb, and object, and adjectives generally precede their noun in English, and certain combinations of tense and person are coordinated. These associations are anticipated and knowledge of them, on the letter, sentence, and paragraph levels, is used to interpret a transmitted message. For example, consider the garbled sentence "nour ango I'll went wi u." Here the spelling errors and omissions are only a minor annoyance. The word meanings can be inferred because of the redundancies of language, and because we know the inherent spelling and grammar restrictions. Note that in this example, the limit of our use of grammatical construction may have been exceeded. The test sentence is partly unintelligible because "I'll went" raises the question of past or future action. Additional content is needed to determine meaning, for example, from the sentence context.

There are also constructions that facilitate the smoothness of our understanding by supplying redundancy but which, therefore, add little in the way of information. Between 5% and 10% of printed words consist of the adjective "the," which is nearly always superfluous. The information content of all these correlations, grammar, and redundancies is not easily quantified but, in principle, they represent a $\Delta S^{(2)}$ that further reduces the novelty value of a message. The quantity $\Delta S = \Delta S^{(1)} + \Delta S^{(2)}$ is information that is stored in the system; in connection with machine intelligence Watanabe[6] calls it "the strength of the structure." This strength, which is necessary for comprehension and interpretation of any message, is part of the interpretive scheme.

Because of this system strength it is estimated that approximately 50% of the letters in a communication can be omitted without greatly compromising the information it carries. This is the basis for many word games. When all these features are taken into account it has been estimated that the remaining information content of English is one to two bits per letter. As the exact value will not greatly affect our discussion in this book, for purposes of that discussion let us take it to be 1.5 bits per letter.

This 1.5 bits is a measure of the variability (or "creative" value) of the appearance of a letter in the context of a message. It is the true information not otherwise included in or deducible from the systematic rules, redundancies, and limitations of the language. It allows for variability and differences of writing style. On the other hand the intervening $4 - 1.5 = 2.5$ bits per letter are not lost. They form the structure on which we build our ability to interpret the message and to locate errors within it. They are the fixed reduction of entropy — the inherent information — that sets the framework within which the variable portion can manifest itself. Both portions are necessary for meaningful communication of information. To the communication engineer the framework need not be transmitted because it is known by sender and recipient of a message, so he does not consider it. To the cryptographer and to the student of language these 2.5 bits, as well as the non-independence value $4.76 - 4.0 = 0.76$ bits, are essential to the structure of language, although they impart different sorts of information.

The main point to be taken from this discussion is that information content measures the departure from maximum entropy, i.e., from complete randomness. It consists of three parts:

1 The contribution due to nonuniform occurrence of individual elements — of the alphabet letters for the case of language. The nonuniformity of occurrence reduces the maximum entropy because some letters are anticipated more than others. This correction makes the entropy level equivalent to that of equally probable elements, the classical entropy level. The universal necessities of language are reflected in the fact that, based on Equation 8-1, the classic entropies of common languages are nearly the same. French, which has different $P_i$ values from English, e.g., $P(e) = 0.17$, also has $S \approx 4$ bits per letter, as does Hebrew, with only 23 symbols (including the space) and nearly no vowels.

2 The "structure" of the system or vehicle being used to transmit messages — for languages this is the spelling and grammar rules and the redundancies. This structure is equivalent to an inherent reduction of entropy. A large inherence allows greater error detection and reliability but reduces the range of variability of the message itself. This structure consists of terms like Equation 8-3, and others at the grammatical construction level.

3 The particular, unique message itself. The type and degree of new information that can be carried by a message are reduced as a result of the uncorrelated distribution [(1) above], and by features of the structure within it resides [(2) above], that is, by the inherent entropy of

information levels of the vehicle. A low inherent information content allows a large "free" message variability. The danger is that if the inherent part becomes too small the completely random messages that result will be meaningless. In the evolution of language there must have been an interplay between these aspects to arrive at a suitable balance, the outcome must have been to maximize the rate and ability to transfer information between human communicators.

It is useful to express these conclusions in the following way; the terms used are displayed on Fig. 8-1, which is adapted from the work of L. Gatlin.[7]

- $S_o$ is the maximum entropy; it occurs when all $P_i$ are equal

$$S_o = \log_2 n \qquad \text{bits} \tag{8-4}$$

where n is the number of elements.

- $I_1$ is the information content of the individual elements. $I_1 = S_o - S$, where S, found from Equation 8-1, is the classical entropy of independent events. $I_1$ is due only to the fact that the basic elements do not all occur with the same probabilities.
- $I_2$ is the sum of information contributions due to rules of conduct of the system (one of which is Equation 8-3). For language these are for spelling and grammar. $I_2$ is the strength of the structure. It is the information that is both explicit and inherent in the rules and

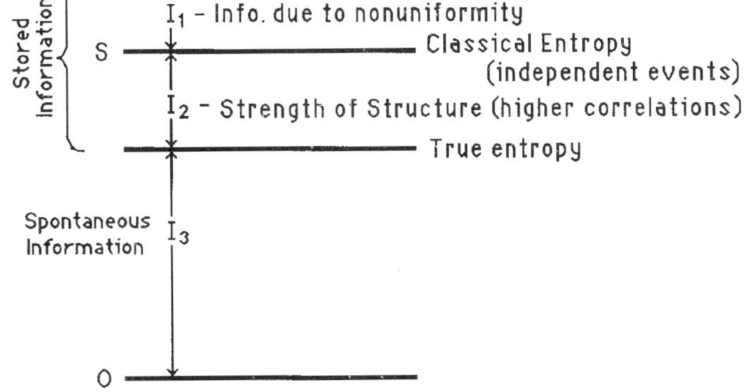


**Figure 8-1** Relation between entropy and information components.

behaviors of the framework of language. The combination $I_1 + I_2$ is the "stored" information of the elements and system.

- $I_3$, the spontaneous information, the primary concern of the communications engineer, is given by

$$I_3 = S_o - (I_1 + I_2) \tag{8-5}$$

## Information in Genetics

The genetic paradigm is that the genes in a cell have coded within themselves the information necessary to build the life form for which they are specific. That life form then reproduces, passing along the genetic plan to a new generation. In Chapter 2 we addressed the question of the appearance of genes and, particularly, the consequences of eucaryotic evolution. In Chapter 3 we described the mechanism of biological evolution through genetic change and Darwinian selection. In the present context we deal with the means by which genetic information is carried.

The basic genetic structure, DNA — deoxyribonucleic acid, consists of two tremendously long chains of organic molecules that are helically wound about each other. Each acts as a support or "backbone" for molecules called "bases," that bond to corresponding bases from the other strand. There are four bases: thymine (T), adenine (A), cytosine (C), guanine (G). While there are no restrictions on the sequence of bases along the backbone, those on one chain can only join with those on the other in the pairs A-T and G-C. As a result each molecular chain is a complement to the other. The DNA molecules have distinctive chemical markers that give opposite directionality to the two chains, and it is the arrangement of A, T, G, C along the molecule that carries the genetic code. Through the meditation of RNA, ribonucleic acid, and various enzyme actions, the DNA can be replicated and can direct the production and distribution of the protein molecules of which living matter is constructed.

Therefore, the genetic language has a four letter alphabet. Since there are 20 different amino acids, a minimum word length of three letters is needed if there is to be at least one unique code word for each amino acid. This unit, called a codon, has $4^3 = 64$ combinations so that the correspondence has redundancy, i.e., more than one combination may lead to the same protein. To transcribe the codon number into an amino acid implies the existence of an interpretive system. Otherwise, for example, calling for number 132 (In base 4 code this is $1 \times 4^2 + 3 \times 4^1 + 2$

$\times\ 4^0$), is meaningless. This interpretation is supplied by the underlying laws of organic chemistry and by stored information in the genetic system in the form of genetic regulatory regions. For example, some of these control initiation and termination of the transcription process whereby RNA copies the DNA sense codon. These and other regions of the gene serve as intermediate instructions, i.e., rules of grammar or the "system strength," if you will, for other and later steps of the process.

Gatlin[7] examined several indices that might be taken to describe genetic information storage. She showed that lower organisms have a wide range of nearly any index she tried. In contrast, vertebrate animals have genetic information that clusters about a value of $I_2/(I_1 + I_2) = 0.7$, meaning that seventy percent of their "stored," or background, information, the interpretive base, is in the form of system grammar. She also found that $(I_1 + I_2)/S_0 = 0.3$, meaning that the potential true signal variability $I_3/S_0 = 70\%$, representing the genetic information that is meaningful code for the particular species and individual differences. This is somewhat misleading because is has been found that this 70% includes a large proportion, estimated to be 25–90% of genetic bases, that appear to have no primary function.

This last statement must be regarded cautiously, especially in view of the great amounts of money and effort that are being expended on the Human Genome Project and the high expectations of its results. This is the vast program to identify each of an estimated 100,000 human genes and to connect it with its influence. That number represents DNA sequences that either code for or influence the production of proteins, e.g., amino acids, RNA, cell enzymes. However these targeted genes represent only about 3% of the total DNA in the human genome. Because the remaining 97% does not encode protein development directly, i.e., it appears to be synonymous or silent, most of it has long been regarded as "junk DNA," indicating that it has no apparent purpose. It has been conjectured that these DNA segments are remnants of earlier instructions whose use has disappeared during the evolutionary process, or that they represent potential future experimental variations, or redundancies built into the system for safety. Alternatively, they may be "accidental" or "freeloading" genes that serve no apparent purpose but which, because they appear to do no harm, have not been eliminated. Recent studies have indicated that a portion of junk DNA actually serves the vital purpose of spacing other genetic parts so that proper overlapping, needed for replication and exchange, can take place during mitosis and meiosis, i.e. it represents genetic grammar and system strength.[5]

Contrasting with these evaluations, is the proposal that junk DNA may carry hidden or indirect genetic tendencies that are not immediately manifest or that we are unable to ascertain, or that they constitute developmental regulatory mechanisms. There is evidence that missing or misplaced lengths of these DNA sections can result in development errors or disease. As a result it appears that, even on completion of the Human Genome Project, estimated to be about 2005, there will still remain a vast unexplored and important genetic domain.[9]

Gatlin's comparison of bacteria, plants, and invertebrates with the vertebrates points up the evolutionary development of an information distribution strategy. There is also a growth of total stored information. A bacterium has of the order of magnitude of $10^6$ base pairs.[10] This amount of information corresponds to only 5000 printed pages (at an average of 1.5 bits per letter) on which to store its genetic blueprints. Long sequences of letters can carry more information and there are about $(10)^{10}$ base nucleotide pairs in a mammalian cell, or ${}_4(10)^{10} \approx {}_{10}0.6(10)^{10}$ base sequences. This corresponds to five million printed pages of information.

Even the tremendous amount of information contained in the genetic code — the equivalent of five million pages of mammalian genetic information — does not seem enough to supply detailed instructions for the making of a complete, complex, advanced animal. For example,[11] "*Homo sapiens* has more nerve cells in his central nervous system than information symbols in his genome." It is more likely that the genetic code contains principles for making plans, rather than the plans themselves. This is a well-known technique. A teacher cannot possibly instruct students how to handle every situation they are likely to meet. Instead, one teaches the principles involved in handling those situations. It has been suggested[12] that the genetic code involves a similar approach. It does not contain inflexible rules of procedure and development but, instead, consists of algorithmic patterns (general rules of procedure) for the response to stimuli. As the situation and cumulative influencing factors change, the algorithms allow cell development to advance in different ways, depending on site in the developing embryo and stage of growth.

In this regard we note that for the bacterium it's mere 5000 pages is adequate because it does not have to produce its own complete lifeform. Instead, for example, parasitic bacteria take control of genetic production mechanism of the foreign cell they invade and use its available capabilities for that purpose.

## Intelligence

The highest achievement of biological evolution is generally taken to be the human brain. Human intelligence results from the brain's great ability to store and process information, a trait attributable to its tremendous complexity. The brain has $10^{10}$ to $10^{11}$ neurons[10] (nerve cells), each of which interact directly with between 10 and 10,000 others, and with even more through secondary contacts. A single stimulus can initiate many thousands of serial and parallel excitations and responses. In the longstanding effort to understand intelligence this complexity has placed the brain functions beyond available analytic techniques. Now, for the first time, the digital computer has allowed simulation or some behaviors attributable to the brain, with the hope of gaining insight into its operation. Conversely, the brain can be regarded as an ultimate computer and its structure has stimulated approaches to computer architecture, with the hope of improving computer ability to solve complex problems.

Attempts at artificial intelligence (AI) in computers have become common in recent years because of the rapid increase of computer speed and capacity.[14] AI is an effort to make the computer self sufficient in evaluating problems and making decisions about their solution. First approaches, which were and still are quite successful, employ logical branches at each of which the computer eliminates one family of possible solutions, based on the input of information and its stored memory of possibilities. By proceeding along the resulting path the machine eventually arrives at a conclusion or at a class or type of recommended actions. This approach uses the sequential, step-by-step instructional procedures that are central to most computer data handling and logic. Its success depends on the detail of stored experience and astuteness of the questions asked at its logic branch points. These are taken from extensive analysis and questioning of human experts so that an AI system is actually a proxy for the systematic knowledge of group experience.

A more recent approach in AI is to develop algorithms that allow the computer versatility in responding to a particular situation with which it is presented. Some of these "smart" machines can handle problems where the available information is insufficient to make firm, logical decisions, perhaps based on probabilities of outcomes stored in its memory. Some of these algorithms include the capability of having the computer alter its own procedures and stored information base, to reflect its ongoing experience in solving problems. These computers can, to a limited degree, learn and change. As mentioned above, with such similarities it is inevitable that AI procedures should be taken as a model for the workings of the human brain, and that the brain should be used as a model for AI development.

One approach is through the recently introduced technique called "fuzzy logic" in solving problems.[15] This acknowledges that, whereas the computer uses strict numerical values to distinguish items, categories, data, etc., the brain deals with property types such as bigger-or-smaller (or big-or-small), louder-or-lower (or loud-or-low), firmer-or-softer, reliable-or-questionable, etc. Although precise numerical distinctions are not usually available in most problems because the relative categories have ill-defined or fuzzy boundaries, the categories themselves can generally be distinguished. By having the computer distinguish between many such fuzzy groups a "consensus" of properties can be arrived at, sufficiently narrowing down the possible choices that a problem solution or answer can be stated with a high degree of confidence. This approach is found to be fast, accurate, and robust (i.e., useful for a wide range of problem types and information variations).

The high order connectivity of neurons has also created the computer research area[16] called "neural networks." Here one investigates the consequences of interconnecting logic elements and computers in local and extended networks. Programs and data are shared in various hierarchical schemes. A related attempt to simulate the brain with computers has generated the subject of "Parallel Distributed Processing" (PDP).[17] In this scheme computations are divided among several computers that perform simultaneously and interact to modify their procedures and results. They may share intermediate results and learn from each other, to arrive at a problem solution. Investigators are developing and simulating schemes[18] to engage in "massively parallel computing," with plans to employ thousands of processing units having extremely high data switching rates, on the order of gigaflops, or even teraflops per seconds. (giga = $10^9$, tera = $10^{12}$, and the "flop" denotes a digital switch, e.g., from 0 to 1, but it is interpreted more generally to include memory query rates, even if there is no actual change of stored information.) In practice, these efforts presently involve only relatively small scale networks and parallelism; investigators use and simulate, a hundred to a few thousand central processor units, at most. However, even with this small number there is developing a theoretical understanding of the use and behavior of interconnected logic systems. For example, it has become clear that there is a probablistic aspect to different outcomes. The uncertainty arises from interactions, learned procedures, and data that are not completely specified. This brings in the relation between entropy and information. Smolensky[19] refers to the equivalent of $I_3$ as "missing information," with the implication that $I_1 + I_2$ are not hidden but are contained in the programmed and learned data and procedures of the AI process.

A considerable effort has gone into the study of intelligent machines. This includes their ability at pattern recognition, i.e., interpreting and analyzing input data, and their ability to control other machines and processes,[17] i.e., to find the right sequences of decisions and precise actions. These steps demand information storage in long and short term memories, and they demand information handling and management; the same abilities we have attribute to living systems. G. N. Saridis[20] writes

> Therefore, the flow of knowledge is the main variable in the function of an intelligent machine.
> – *Knowledge* is the function of removing of ignorance or uncertainty in the operation of an intelligent machine and may be measured by entropy, which is a measure of uncertainty.
> – *The Rate of Knowledge* is related to the flow of knowledge in the machine with its purpose being the reduction of uncertainty and it is a measure of its Intelligence.

Saridis' definitions are not the only possibilities but his references to entropy and uncertainty are generally accepted. More pointedly, he states that "The theory of intelligent machines may be postulated ... such that it minimizes its total entropy." He demonstrated this for control processes and Smolensky has demonstrated it for pattern recognition processes. While entropy reduction has been regarded as a property of living systems we see that it continues in the most current of efforts to extend our reach beyond the constraints of nature.

## The Evolutionary Parameter

The discussions in this and the previous chapter have tried to demonstrate some of the relations between degeneracy, entropy, and information. These concepts are pertinent to all classes of things, from the simplest atomic systems, through the stages of life, to the most advanced "thinking" machines, natural and artificial.

Theoretical biologists have long recognized that the evolutionary process of life can be viewed in terms of information. Consider the instant when the very first DNA molecule (or its inorganic or organic predecessor) was duplicated and led to others like the original. That DNA held and had passed along the rules for its own replication and, perhaps, for the production of the cell whose structures protect and manage its existence and propagation. It was a set of instructions and its propagation involved an important communication of information. M. Eigen, in his discussion of *Steps Toward Life*, states[11]

> The transition from inanimate to living structures took place with the increasing ability to wield information in a quasi-intelligent way. The first steps towards life were taken in a chemically rich environment, one in which information-storing molecular replicators appeared. Only these had the capacity for optimizing, and thus for a teleonomic approach towards goal-determined behavior. They preserve what has been attained, replacing it only with what is more expedient.

H. R. Branson[21] goes further, adding

> It seems increasingly plausible to assume that information should be considered as the basic parameter for describing living systems …

J. A. Wilson[22] expresses the extension of this concept by observing that

> Natural selection brings information into the gene pools of species. This leads to increased complexity, better fit of populations into their niches, increased population numbers, and increased genetic variability.

J. S. Wicken[23] agrees by pointing out that

> All natural systems are in the business of acquiring and using information through ecological interactions-the ecological arena being broadly construed to include the mix of physical, biological, and societal ingredients that decide whether a given innovation is successful. Separating the introduction of a novelty from its selection was Darwin's fundamental insight. What works ecologically is that which is incorporated as information.

These are but a few quotations from the literature that relate and equate information and evolving systems.[24] From this viewpoint, the story of biological evolution describes the growth of information generation, storage, and processing abilities, and their use to produce ever more complex creatures that are successful in enhancing and communicating their abilities to future generations. It is simplest, on the genetic level, to consider only the stored information in the coded nuclear DNA. However we should not separate this from the ability of evolving life forms to process information, and from the human faculty for creating information. Advanced evolutionary stages are not only more complex in themselves, than lower forms, but they can also store, process, and develop more varied forms and greater amounts of information.

Before continuing, we should make a point concerning terminology. From the correspondence of their mathematical forms, as given by Equations 8-1 and 8-2, and from Figure 8-1, we recognize that information is a form of entropy or, more accurately, it is the complement of entropy. However entropy is basically a thermodynamic term, associated with thermal systems and heat energy transfer. This leads to some awkward and confusing constructs when it is applied to living systems. For example, to describe the fact that such systems function to increase their ordering, which lowers their entropy, the energy transfers involved are said to be converted to negative entropy. Alternatively it is claimed that they generate "negentropy," or that they export entropy.

These connotations are convenient and have relevance and meaning for the physicist wishing to calculate quantitative entropy-information changes due to specific physical processes, but they are not truly proper. Entropy is a property of a system, resulting from its probability of occurrence, i.e., from its microstate degeneracy. Heat and energy may be transported, — the negative being associated with direction of energy flow — and the system entropy may change as a result, when its properties and microstate probabilities change. But, strictly, entropy does not "move"! Furthermore, for the most part our usage is divorced from these thermal aspects, though they influence and possibly control some of the changes we are considering.

In addition, to refer to the evolutionary parameter as entropy frequently leads to confusion when referring to an increase of some property of information which might cause a reduction of an aspect of entropy. We will therefore drop the unit conversion factor $k'$, in Equation (8.2), and use the more biologically relevant term information, when discussing most features of evolution.

Perhaps the most striking aspect of the Evolutionary Trajectory is its continuity, from the earliest cosmic moments, through the biologic life processes, through the occurrence of intelligence, to the products of that intelligence. Interpretation as a series of logistic escalations implies that its regularity is through modifications of some variable, and the specific nature of events 8 through 12, along with the considerations given above, leave little doubt that the pertinent parameter is informational. We, being the generators of the latest events, can understand their underlying commonalty more clearly than that of earlier events, but the Trajectory itself dictates that this parameter must be pervasive.

Our conclusion, therefore, is that the Evolutionary Trajectory of Figure 6-7 consists of a series of logistic escalations of information. Consistent with Price's condition that each escalation involves "a slight change of definition of the thing that is being measured," different aspects of this multidimensional variable are likely to be prominent during different evolutionary escalations. These include its growth, organization, use, generation, storage, processing, and communication.

An appreciation of these differences is necessary to understand the widely different estimates that have been made of information content in systems. Henry Quastler[10] discussed the results of investigators who have calculated the entropy and information of generating the organic molecules of a bacterium from their precursor constituents. They find an upper limit of $10^{10}$ to $10^{13}$ bits. This is a large range that, even so, neglects the incorporation of those molecules into the cellular structure, not to mention the information needed for cell dynamics. From another viewpoint or, more accurately, on another comparative basis, he points out that the genetic information of the bacterium is only about $10^5$ bits. This indicates the great problem of comparing informationality across evolutionary escalations.

## The Elaboration of Evolutionary Information

Küppers[25] casts the varied aspects of information into an alternative grouping when he writes:

> The concept of information that we have introduced possesses three dimensions:
>
> a. the syntactic dimension, which comprises relationships between the individual characters.
> b. the semantic dimension, which comprises relationships between the individual characters and what they stand for, and

c. the pragmatic dimension, which comprises relationships between the individual characters, what they stand for, and what action this implies for the sender and the recipient.

Syntactic information is that given by Equation 8-1; it is the engineers' informational entropy, corresponding to S on Figure 8-1. For such information to have meaning there must be a semantic agreement of its interpretation. For example, there could be no realization of digital computation without the underlying bases of electronic signal processing, circuit fabrication, theory of languages, etc.; these are parts of the informational basis of the digital computer. In a similar way we pointed out earlier that calling for genetic codon 132 has no meaning or effect unless it can be semantically understood. There must be an interpretive level, here consisting of genetic regulatory regions and the laws of macromolecular chemistry, that recognize the code and transform it into a particular form so that its structure and/or function can be developed. This interpretive process represents additional information that might either enrich, e.g., by adding allowed alternative meanings of the codon message, or limit its possible realizations, e.g., by imposing interpretive rules and controls.

Once a syntactic message has been interpreted by the semantic mechanism, along with any possible variations or errors this may imply, Kuppers indicates that a pragmatic dimension must be considered for the realization of that interpreted message as a final structure or action. The semantic and pragmatic dimensions can both be considered as "interpretive;" the former corresponds to $I_2$ on Figure 8-1. The distinction between them is worth pursuing.

For this let us consider that an intelligent person has heard and understood a verbal communication. In terms of our discussion, above, this is a syntactic message received by a recipient, whose semantic system strength interprets it. Any physical, intellectual, or emotive response or attitude is a result of interactions between this message, other available information, and his or her complex response system. These pragmatic modes and influences are separate from the direct informational process alone. They are, perhaps, much more difficult to understand than the syntactic and semantic aspects, but it is misdirected, as some have done,[26] to attribute to them an interpretive function *in the informational sense*. Surely they use information in the "pragmatic dimension," but that utilization is a reaction rather than an interpretation. Pragmatic differences of response to the interpreted message constitute part of the variability of individuals at all informational levels. They contribute to differences in the ontogenetic orthogeny of a mammalian fetus as well as to the generation of differences between unicellular creatures.

When the first nucleic acids were formed their construction was based on the underlying interpretive level of molecular chemistry. The cell is formed from genetic information based on the interpretive level of the interactions of organic macromolecules. Going further, it can be assumed that each escalation on the evolutionary trajectory is built on the informational level of the preceding stages. When a genetic codon calls for generation of one of the 20 protein molecules that make up living matter, there must be a system with sufficient strength, likely tRNA and enzymes, to interpret the command. And, on a higher level, when the Hominids evolved, eventually leading to Homo sapiens, the elements of change were built on the existing interpretive schemes (semantic and pragmatic) of the Hominoid superfamily from which it derived. In that case the Hylobatids (gibbons) and Pongids (great apes) build different variations on the same system base. This is implied when it is stated[27] that part of the uniqueness of living systems lies in their dependence on cladistic history. The dependence on distant and immediate history means, among other things, that the coded messages is not independent of the milieu in which it is decoded; the genetic unfolding process is not single valued either within a species or, certainly, among species. In part this explains why geneticists find that some common codes and combinations can be found in widely different species, families, and even phyla.

Most genetic informational estimates neglect aspects that are not simply included in "counting" schemes. This can be illustrated by an example from the genetic code. It is simplest to consider only the stored information in the coded genetic DNA but it is easily seen that this alone is insufficient. The three-dimensionality of a protein structure carries a great deal more information than does its one dimensional arrangement of DNA code. The transcription process must supply this "added value." There may be several sources of these extra factors. For example, the estimates made by Gatlin,[7] referred to earlier, result from analysis of the static system, be it macromolecular or genome. However the semantic and pragmatic interpretive dimensions contain dynamic portions as well and, as discussed in Chapter 3, pragmatic realization depends on the influences of changing endogenous and exogenous factors and of other chromosomal regions.

Modern molecular genetics has disclosed a number of interaction mechanisms. Structural genes contain "introns" that may shuffle the protein domains coded for by exons; cis-acting and trans-acting genes control the expression of structural and other regulatory genes at local and distant positions; homeoboxes exert major controls on gene expression; gene processing enzymes respond to signals from operons and repressors, and these and the various mechanisms that repair gene damage are sensitive to environmental influences.[28] Campbell[29] has pointed out that

> Multigene family organization enormously increases the capacity of DNA to store, process, and express genetic information. It allows-and demands-evolution to proceed quite differently than from simple, single-copy genes.

As there are four base types each genetic base carries an information content given by Equation (8-4) as: $S = \log_2 4$. For an array of N equally likely bases, each acting alone, the maximum information is $I^{[1]} = N\log_2 4$. However, if the elements, in addition, interact in contiguous groups, the sum over all family sizes gives the information capacity,[30]

$$I(\max) = \Sigma\, I^{[i]} = (1/2)\, N(N + 1) \log_2 4 \approx N^2 \qquad (8\text{-}6)$$

where the approximate equality is for large values of N. Thus, for the viral value quoted above ($N \approx 10^5$) we find that $I(\max) \approx 10^{10}$ bits whereas $I^{[1]}$ is only $2 \times 10^5$ bits. Even this great expansion of possibilities does not indicate the full potential of genetic information since, as pointed out above, interactions also occur between non-contiguous genetic regions, some quite distant and some relatively mobile. Therefore the growth of biological information, described by the evolutionary trajectory, might arise from increased storage size (i.e., genomic length), from a redistribution of semantic and spontaneous information or, probably more effectively, through allowed interactions.

These changes are manifested through the cyclic escalations of the Evolutionary Trajectory, each of which has a different effect. For example, during the cycle following Big Bang we have to deal with thermodynamic heat and local reduction of entropy through particle and material ordering. This seems to have been followed by an escalation in which organic information storage ability developed, and then a cycle or cycles when storage efficiency increased, when physical complexity developed, when information utilization grew, and/or when information collection became significant. Perhaps there were other intervals of information processing ability, an escalation involving capability of information generation, and the most recent stages surely appear to have involved information modification and information transmission.

The changes result in the increase of life-form complexity and development of the brain, culminating in the human cerebrum. They are continued by the appearance of civilization, and continued through writing and then printing and, most recently, by the electronic computer. Contrary to most belief, major evolutionary variations did not cease when *Homo sapiens sapiens* came on the scene. It has taken a previously unrecognized direction, while still maintaining its cybernetic trajectory and cyclic nature. In view of this we have no reason to suspect that it has ended. This will be discussed in Chapter 10.

## Are There Alternative Trajectories?

Having demonstrated information to be the hidden parameter of the Evolutionary Trajectory we must address some criticisms that have been made about that parameter and about the choice of trajectory events presented here.

We have already mentioned the problem that many important developments are omitted from Figure 6-7. The physicist may well wonder how it could omit the important moment of separation of radiation and particles, or of the formation of galaxies. The biologist will surely be skeptical about the tremendous leap from the initiation of Metazoan evolution, with the eucaryotic radiation of event number 2, to the appearance of mammals. This seeming neglect of the major portion of organic evolution appears inexcusable. For example, the appearance of vertebrates, with their unique body structure, or the amphibian/reptile attainment of independence from an aqueous environment, seem to be major transitions. And the historian can surely indicate momentous events in the development of our civilization, such as the invention of the wheel and axle, the merging of Greek and Persian proclivities under Alexander the Great, or the first global circumnavigation, none of which are manifest in the evolutionary trajectory.

Our response to these objections must point out that the quantity "information" is neither a purely astrophysical, nor a biological, nor a social/historical one. We should not expect *exact* agreement with criteria from other fields, although the selection of all our critical transition points has been based on criteria and category divisions previously established by specialist in their own areas of study. For example, the events on Table 6-3 are the standard biological-phylogenetic categories used in text books to describe the cladistic descent and structure of mankind. It then appeared, from the trajectory itself, that event $\alpha$ on that table does not conform to the sequence of the other events.

It has been our intent to avoid arbitrarily or subjectively chosen points to conform to a preconceived connectedness. The objections, that one or another important point has been overlooked, primarily reflect the subjective evaluation of the protester. For example, it has been claimed that the major technical advances made by society have been by learning how to make and use new materials. This is illustrated by the advent of the iron age some 4000 BP, by the use of fossil fuels for energy, by the discovery of antibiotics, of plastics, of semiconductor materials in electronics. For this reason it is reasonable that the trajectory should reflect the first use of copper or the discovery of alloying to make bronze, as the first case of this technological mastery of nature's materials. Although this was indeed a significant metallurgical turning point, with informational content, it does not appear to constitute a critical, evolutionary transition of information. Our use of criteria and divisions established independent of our usage here, and accepted by pertinent experts, avoids this question of subjectivity.

In recent years understanding has grown of the relations between evolutionary events. In addition, the concepts of General Systems Theory have been developed and applied, and we have begun to appreciate the fundamental mechanisms of change of systems, in general, and organic systems in particular. As a result, several students of evolution have attempted to divide its progress according to specific criteria and chains of events that seem to them to be primary departure points for change. Thus, the paleontologist J. William Schopf has edited a book entitled *Major Events in The History of Life*,[31] in which the chapters discuss the history of:

The Origin of Life [chemical origins]
The Oldest Fossils [Procaryotes to Eucaryotes]
The Earliest Animals [Metazoans]
Land Plants
Vertebrates
Mankind

Taking the vantage point of the growth of complexity, within the framework of general systems analysis, John Maynard Smith and Eors Szathmary[32] have written *The Major Transitions in Evolution*. Their choices, which emphasize changes in the transfer of information, are:

Replicating molecules
Independent replicators [chromosomes]

RNA and enzymes [genetic code]
Eucaryotes
Sexual reproduction
Cell differentiation [animals and plants]
Colonies of individuals [cellular and social]
Human societies [language]

Klaus Haefner[33] considers six levels of the handling of information in evolution:

In physical structures and systems
Molecular genetics
Neural information [handling external information]
Social systems
Technical processing
Sociotechnical

Perhaps the broadest view has been taken by Alexis V. Jdanko, who has published several papers on an evolutionary-entropic interpretation of information.[34] In particular, he chooses[35] the "main phases of evolution" to be:

The Universe [Big Bang]
Solar System [formation of earth]
Life
Hominids
Homo sapiens
Civilization
Modern era [beginning 400 years ago]
Computer

and in a discussion of cognition mechanisms[36] he lists:

Speech
Writing
Printing

All of these divisions are based on the analyses of perceptive workers in the study of evolution. Their choices are insightful and in each case the writers base their selections on a particular set of criteria, in agreement with the general understandings that are extant in their particular technical areas. Their selections of events are guided by this specialists point of view so that, while the criteria are appropriate basis for analysis, they cannot help but reflect a narrowed viewpoint on the totality of evolution. The degree of agreement between the listings indicates that all their views have validity as parts of the complete representation. The qualitative relations and judgements on which they are based are not definitive enough to allow differentiations to be made between events of major and of minor significance, or to indicate whether their listings are complete.

The events presented here, on the Evolutionary Trajectory on Figure 6-7, were also selected by experts. In this case, however, their goals were not to conform to an evolutionary preconception but only to categorize the characteristics of their own specialized fields of study. By putting their results in time sequence we were able to distinguish major and minor events because, for the first time, we can be guided by quantitative theory, that of Logistic Emergence. While other and intermediate, events may, indeed, be of great significance they are not critical transitions of the greater, *informational* development of the Trajectory. As we discuss in the next chapter, it is difficult to ascertain just what informational property was developed during each cycle or by how much it has changed but each step marks the end of one cycle and the beginning of the next. Each transition subsumes the intermediate taxonomic changes and informational attainments up to that point, and then introduces one or several new modes of development.

## Additional Considerations

Even with the advantage just mentioned it may be felt that the choice of transition events is rather arbitrary because the logarithmic scale reduces deviations from the straight line relation, so than nearly any selection of events might be made to conform to some trajectory or other. Here we point out that experience indicates otherwise. The clear nonconformity of point $\alpha$, on Figure 6-1 is a case in point. Even beyond this, the necessity to displace one event on Figure 6-3, indicates the selective feature of these trajectories and, in that case it directed attention to on otherwise neglected feature of the religious trajectory. We should also mention that consistency of the trajectory was significant enough to demonstrate the need for an adjustment of the time reference of Figures 6-6 and 6-7. It is one thing to claim that any selection can be made to conform to a reasonable trajectory and quite another to generate such a meaningful alternative.

The Evolutionary Trajectory exists over eight orders of magnitude of time and displays thirteen data points. This imparts an undeniable reality to its existence and significance, and leads to an *a posteriori* reasoning about the fact that some events are on the evolutionary trajectory while others, that might be expected, are not. By this is meant that the existence of and continuity of the final trajectory implies an internal consistency of its points that is not a property of others. The differences may be of magnitude or of type. Since we do not know the precise nature of the evolutionary parameter at each escalation we must not impose our *a priori* attitudes, other than in a general way. In part it is continuity of the trajectory that defines the transitions of which it is constituted.

As an example of this, it was mentioned above that biologists would expect the appearance of vertebrates to be transition point. Not only does it indicate the development of a new, fundamental body plan but Gatlin[7] has demonstrated a changed genetic storage strategy at this evolutionary point. It appears, however, that this was not *the* critical change to appear on the trajectory. With regard to what may be the more substantive change of information usage in mammals, Barlow[37] states that

> Embryonic development in mammals is distinct from that in other vertebrates because it depends on a small number of imprinted genes that are specifically expressed from either the maternal or paternal genome. Why mammals are uniquely dependent on sexual reproduction and how this dependency is dictated at a molecular level are questions that [are being] intensively investigated …

Furthermore, in considering the appearance of mammals it has been pointed out[38] that

> The neocortex is a highly developed part of the brain's cortex — the outside layer of gray matter — that began its evolution with mammals. It is thought to be the region where memories are finally stored and where reasoning takes place.

Most of us are used to consider linear time intervals and may therefore question the appropriateness of making an early evolutionary cycle, that may have lasted for tens of millions of years, equivalent to a later cycle, which may have been only millennia long. After all, the time intervals are so tremendous in the early escalations of the trajectory that many important developments were able to occur and our representation is forced to neglect them. However the trajectory seems to dictate that the absolute length of interval may not be the best measure of elapsed time; the logarithm of time is its coordinate and this makes the hundreds or thousands of years of the most recent cycles developmentally equivalent (or informationally homologous) to the millions of years of the first few cycles. This is a new way if viewing the passage of evolutionary time but, nonetheless, it seems to be appropriate. For example, *Homo erectus* survived from $1.6(10)^6$ BP to about $0.5(10)^6$ BP. The difference of the logarithms of these dates is 0.51 logyr. By comparison, the dinosauria, which are particularly noted for their longevity, reigned from roughly $2.3(10)^8$ BP to $0.65(10)^8$ BP, a difference of 0.55 logyr, not so much greater than H. erectus.

This reduction of elapsed times from years to, say, fractional escalation cycle lengths seems to be reflected in the reduced significance given to the informational changes that accompany the otherwise very important intermediate morphological and physiological changes. In addition it emphasizes the fact that the most recent few cycles are developments by *H. sapiens sapiens* alone. Many ecologists and biologists reject any seeming emphasis on mankind over simpler lifeforms. After all, algae or insects, are certainly more prominent, by nearly any quantitative measure, than human beings. And the sustainability of life on earth seems to be more critically dependent on them. However the Evolutionary Trajectory is a representation of informational change, and by this measure there is no doubt about the place of humankind.

Another item that needs comment deals with the extent that our results depend on the exact forms of the logistic equations, 5-4 and 6-1. The ordinary logistic Equation 5-4 has been widely used to describe both simple and vastly complex phenomena. Despite its simple analytic form, involving only 2 parameter, it seems to encapsulate the essence of many real developmental processes, i.e., it cuts through obfuscating details of behavior that might be expected to produce deviations from its simple form. For example, processes that begin more rapidly than the slow, initial exponential behavior, but that eventually saturate, can also be fairly well described by the simple logistic relation. The same also appears to be true of the emergent logistic Equation 6-1. It may not be necessary to have exactly logistic constituent terms for the trajectory of Equation 6-3 to appear. The varied examples given in chapter 6 seem to indicate that the simple semilogarithmic relation is more robust than is indicated by the special case from which it is derived.

We must be careful not to misinterpret one particular apparent implication of the trajectory. The growth of information along the evolutionary trajectory might be interpreted to imply that, to counterphrase Stephen J. Gould from our introduction, the history of life (and in our expanded understanding all evolution) does "have an arrow specified by some vectorial property ..." This creates a problem because of the strong, independent, and universally accepted evidence that, at any instant, evolution proceeds in a random way. Its apparent direction is merely the path resulting from these random excursions. It is very important to point out that it is a mistake to infer that informational growth drives the evolutionary process. In the next chapter we will show that the enhancement of information is also a result of random changes and that, like its counterpart, entropy, information is a measure of change rather than its motivation.

In Chapter 9 we also discuss the relation of complexity to evolution. Among those studying organic evolution it is widely held that increasing complexity is the critical parameter of evolution's advance, as the means by which the true, local and instantaneous forces manifest themselves.

# 9: PROGRESS IN TIME

## Complexity and Information

In Chapter 6 we found the Evolutionary Trajectory of and on the earth. This expanded the idea and continuity of evolution beyond the accepted domain of solely organic change, to include the cosmic, on one end of the time scale, and the societal and technological on the other. While the quantitative relation of this trajectory is new, the general idea of progressive development, particularly among the species, has been a topic of conjecture since the Greek philosophers. It certainly became an active area of investigation following the precisely documented evidence given in Darwin's "Origin of Species" in 1859. Lacking reason to suspect otherwise it has been generally assumed that evolution is limited to the changes of biological creatures. In light of the Trajectory, however, we see that it still an active process though, as stated earlier, it has taken a different, unexpected direction.

In Chapter 8, we were able to infer a progression of information as the logistic evolutionary parameter. Undoubtedly, this is consistent with its long recognized importance, but it presents something of a difference from General Systems Theories of Hierarchical Organization where a quantity "complexity" is generally taken to be *the* appropriate evolutionary value function. Superficially the concept of complexity seems clear enough but it is difficult to define precisely, though attempts are being made to do so. In discussing *The Emergence of Genetic Organization*, Prof. C. R. Woese[1] considered the relation between complexity and information. He states:

> The first problem is to define "complexity". Clearly, equating it to "information content" is insufficient — particularly if information content itself is defined in the usual way (e.g., assigning 4.3 bits to every amino acid in a polypeptide)... Yet complexity is a function of information content in that the latter places an upper bound on the former. Complexity is more akin to that elusive quality, information "meaning" — something that is a function of the nature of the system for which the entity is information, but something that we cannot yet define well enough to quantitate...

In a similar discussion K. Haefner[2] points out that

> All systems practice self-organization by means of information processing over and over again using internal information as well as the steady influx of external information. Thus, ... the whole universe as well as all of its "subsystem" can be understood as information processing systems.

Therefore the growth of information and of complexity are related. We now recognize that a part of the definition difficulty dealt with by Woese, and others, lies in the assumption that a single definition is sufficient. His reference to information "meaning" is, likely, closely related to the similarly varying semantic dimension, $I_2$, discussed in Chapter 8. Price's criterion for escalation indicates that the uniqueness of either is unlikely over the entire course of evolution.

The events that constitute the Evolutionary Trajectory vary widely in nature, ranging from the astronomical to the conceptual, and include the biological and the inventions of mankind. This implies that complexity, unless it is interpreted in an unusual way, is limited in appropriateness and range of applicability as the overall evolutionary value function. In the quote given above, Woese indicates that complexity is a subset of information, "in that the latter places an upper bound on the former." He states, further that "complexity is ... something that is a function of the nature of the system ...." A major part of this system nature is what constitutes the physical and physiological details of a taxon, so that it is appropriate to describe the growth of complexity by the progressive history of taxa. Therefore, the geologic record of dominant taxa on the earth can be taken to demonstrate the growth of complexity or, more exactly, of "taxonomic complexity;" that record is just what is given by Figure 1-3. Figure 1-3, then, is the trajectory of taxonomic information, or of organic complexity; it represents only a part of the total informational development.

Figure 9-1 shows both the informational, Evolutionary Trajectory of Figure 6-7 and the complexity trajectory of Figure 1-3. The latter falls above that of total information; its cycles have shorter periods so it develops more rapidly. It has ended after its event 10. The universal nature of Information, perhaps deriving from its relation to entropy, seems to allow it to continue.

## Entropy and Information Growth

D. R. Brooks and E. O. Wiley[3] have proposed definitions of Organization and Complexity as part of a theory of biological evolution that

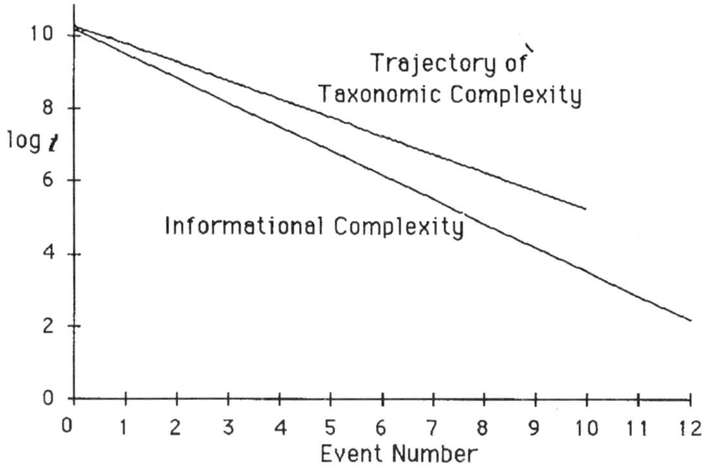


**Figure 9-1** Comparison of trajectories.

appears to be mostly in agreement with the concepts and relations presented here. They show that the growth of information is consistent with both species and cladistic evolution, under the assumption that both $S_0$ and $I_3$, of Figure 8-1, are increasing through geologic time. The difference $S_o–I_3 = I_1 + I_2$, the inherent system strength, is defined by them as "Organization," and $I_3$, the individual variability, is taken as "Complexity." While their results do not depend on the particular form of growth of $S_o$ and $I_3$, they illustrate entropic growth as shown in Figure 9-2. Here the lower curve represents $I_3$. In view of the schematic nature of this figure we ignore the fact that while the growth of $S_o$ and $I_3$ may be monotonic it is unlikely that they are smooth, being subject to the escalation processes.

Of greater interest is the location of the origin of the time axis, which they place at the moment of replicating, lifelike forms, whereas the Evolutionary Trajectory indicates that the growth of information begins at least as early as the birth of the universe itself. While this continuity between the cosmic and biological may be surprising it is, in fact, already known[4] that the Big Bang theory leads to an entropic behavior very much like Figure 9-2.

Rapid expansion resulting from the Big-Bang has acted as a source of increasing $S_o$, the maximum possible cosmic entropy. This occurs through two mechanisms: First, because the available volume of space is growing the number of subvolumes increases and, along with it, the total number

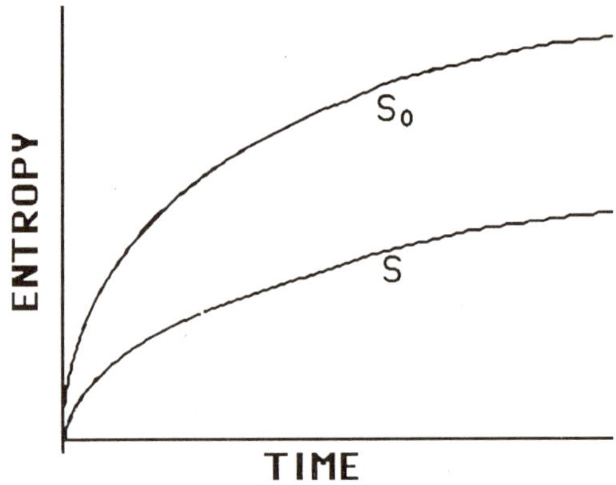


**Figure 9-2** Proposed evolution of entropy components.

of different configurations available to the particles it contains. This gives larger values of W, the potential degeneracy in Equation 7-2, and therefore of $S_o$. Second, as early expansion proceeded the cosmic temperature dropped until photons and nucleons (protons and neutrons) could precipitate out of the initial energy content. Subsequent nucleosynthesis activity lead to the formation of higher elements, with the release of energy and accompanying growth of chemical entropy.

For both of these mechanisms the actual entropy, S, illustrated by the lower curve of Figure 9-2, was less than its maximum possible value. In the first case, gravitational forces restrained the expansion of matter, limiting the attained volume of space and, thereby, causing $S_o$–S, the "organization" of stored information, to increase. In the second the lower density of particles in the expanded universe, and the further reduced temperature, lowered nucleon interaction rates so that the realized chemical entropy growth was less than its potential value. As a result $S_o$–S again experienced an increase.

We have already discussed that these entropy reductions are equivalent to information increases so that continuity of the evolutionary trajectory at events 1 and 2, the cosmic to life transitions, can rightly be regarded as resulting from related, ongoing processes. The detailed mechanisms may differ but the commonality remains. As Layzer[4] points out: "The ultimate cause of order in the universe is the cosmic expansion."

## The Dimension of Time

Consideration of the nature of time is an ancient and sophisticated activity. The psychological and physiological effects and evaluation of the passage of time are the stuff of philosophers, poets, and physicians.[5] The physicist's view begins with Newtonian mechanics, where time and length are the basic, unvarying dimensions necessary to describe existence and change. However, when viewed from the aspect of the four dimensions of the universe: length, width, height, time, an asymmetry is immediately seen. While we can move forward or backward in the three spatial dimensions we only move forward in time; our experience tells us that the universe is causal since past events lead to, or generate, future ones. (The acceptance of precausality was not always common.[6] The once popular teleological philosophy of "final causes" held with a postcausality. There the future attainment and perfection of all things, i.e., of their positions, forms, etc., determine the changes they must undergo and the actions they must perform. This is rejected today.)

In view of this anomalous property of time in our understanding of nature, it is strange that the microscopic dynamical laws of behavior, as conceived by physicists today are, nonetheless, symmetric in time. In a formal way one can describe the situation where time runs in reverse so that changes are determined by their future condition. Such reversibility is especially true on a microscopic level where fundamental particle interactions are indistinguishable in the forward and reverse directions. Two colliding atoms or billiard balls rebound in the same way whether time change is positive or negative or, as discussed in Chapter 8, whether the film sequence is shown forward or reversed.

However, in Chapter 7 we discussed the fact that macroscopic behavior brings in statistical likelihood and with it the idea of entropy. Its probablistic property produces unacceptable behavior when time is reversed, e.g., billiard balls bounding out of pockets or children "preceding" parents. It imparts direction to the time sequence of events so we cannot "remember" the future. What we do remember is the reality of experience and history as the prime manifestation to the traversal of time. This history is also seen in the evidence it generates, such as the occurrence of fossils and the phenomena of ontogeny and of evolution. One interpretation of this considers the distinguishing feature to be the particular initial conditions of our universe — the point and direction at which things start. In essence this conforms to our experience that the universe is causal, requiring that past events lead to future ones. What we subjectively experience is difficult to pin down but as "sapient sapient" creatures we are aware of the passage of time; we are conscious of temporal change.

We have already mentioned, in Chapter 2, that time is a derived property in Einstein's Theory of Relativity, and in the absence of mass it cannot be defined. This is the basis of the statement that before the Big Bang, i.e., before there was mass, there could be no time, so that t = 0 at the Big Bang. In spite of this it is very tempting to extend the degeneracy-time, or entropy-logtime relation that is the Evolutionary Trajectory, back beyond point zero to an earlier, negative time period. When this is done we find that an evolutionary transition point (–1) would occur at about $7(10)^{10}$ BP. Whether such an extrapolation has any meaning is purely speculative. This is not to say that there has been no such speculation on conditions preceding or, rather, at the instant of the Big Bang.[7] Some theories that illuminate the first instants of our universe have been mathematically extrapolated to describe possible states from which those first instants developed. Such analyses are recent and highly controversial; perhaps further study of their meanings, particularly as guided by new measurements of the cosmos, will yield more definitive results.

Our *perception* of the directional flow of time has been related by some[8] to the physical nature of our expanding universe and, specifically, to the cosmic increase of entropy, which is the only truly time-directional physical parameter. According to one dynamical scenario the universe may eventually reach a maximum physical extent and then begin to contract. This raises the question, that has received diverse answers, of how and if this would affect our perception of time. In view of this question it is interesting to speculate on another feature of the Evolutionary Trajectory. We have interpreted it to mean that evolution is a process under which aspects of information increase and degeneracy decreases as time advances. This statement implicitly assumes our Newtonian, experiential belief that time is fundamental and independent and changes uniformly.

However, the functional relation of the Evolutionary Trajectory can be inverted. This would imply that time is the dependent parameter while information or degeneracy is the independent variable that changes uniformly. We might then conclude that our perception of time is somehow a result of experiencing the continual process of degeneracy and entropy reduction in our corner of the universe. Like the other examples discussed in Chapter 1, the semi-logarithmic relation would then be but one other example of adaptively extending our range of appreciation, now of time. The mechanism of such an experiential comprehension is no simpler to understand than that relating time to an expanding universe. Does it imply that we would perceive time as changing in the reverse direction in outer regions of the universe where entropy is increasing, or would other local ordering effects still dominate? Like the cosmic expansion it may put limits on our sense of time. In the next chapter we will show that there appears to be an immanent limit to the present evolutionary process. Therefore we may well ask whether our view of time and process rates will also change if the evolutionary information change alters it nature or its manner or rate of change. As with other philosophical and physical — perceptional questions that arise from our discussion, we leave such consideration to another forum.

## Measures of Information

It has already been noted that no quantitative comparisons exist for changes of entropy or information between the various evolutionary steps. One contribution to the resolution of this problem is obtained from the Evolutionary Trajectory itself. Recall that the magnitudes of successive escalations are related by the quantity $R_k = \eta_{k+1}/\eta_k$. The detailed form of an escalation is described by the power n, and if nR changes slowly the slope of the trajectory is -nlogR. This slope was found to be –2/3 so that if we take n = 1, its most likely value, we have R = 4.64. This is only moderately useful because of the lack of clear understanding of what common or different aspects of information need to be included at each stage. Price's "change of definition of the thing that is being measured" implies different physical and descriptive parameters of successive emergent stages. Any comparison would be between disparate aspects of information.

Thus, it would be meaningless to state, for example, that the information at the end of escalation 4 is $R^3$ greater than escalation 1. In Chapter 8 we observed that even a partial solution of the simplest case defies estimation in other than a general way; investigators who calculate information content of simple biological systems, from varied reference levels, arrive at widely different results.

Therefore, rather than seeking total information, we might approach the Evolutionary Trajectory by asking what aspects of information are being developed in each cycle of the process of evolution. Consider the information content of a digital computer; by this we can mean any or all of the elements of the partial listing on Table 9-1.

What part of this listing places the digital computer on the evolutionary trajectory? The design instructions (genotype) would seem to encompass all the information of the hardware (phenotype). Perhaps it overstates the meaningful system information? Since both design and hardware restrain the type of software that can be employed, are the entire contents of the software a separate informational item?

TABLE 9-1
INFORMATION CONTENT OF A DIGITAL COMPUTER

- SOFTWARE (INTELLIGENCE)
  The specific program(s) being performed at any time
  The collection of programs stored for future use
  The operating system code for instruction and data manipulation
  Stored data and processed manuscripts

- HARDWARE (PHENOTYPE)
  The arrangement of electronic components
    layout and composition on integrated circuit chips
    printed circuit board realization
    microelectronic interconnections
    placement and cabling of subsystems

  Communications means
    output — display monitor, modem, audio coils
    input — keyboard, thermal sensors, scanners, touch zones

  Power supply

  Accessories

- DESIGN INSTRUCTIONS (GENOTYPE)
  Plans and facilities for fabricating the hardware in a consistent way, both on microscopic and macroscopic levels
  Facilities and directions for installing operating system instructions
  Selecting appropriate software to accommodate to hardware
  Placing limits on, and degrees of freedom of operating system

A different approach to this issue raises the question: When certain fabrication techniques are standard and always available, must the measure of information include them in each generation of computer? For example, if certain production methods and/or integrated circuit elements are universal in the plans and hardware listing, need they be specified and counted as additional information of subsequent computer developments? What if specific coding and manipulation are so common that their use does not really constitute new or useful information in the software specifications?

These questions can be extended to biological systems. Geneticists have found that the transcription of genetically stored information, to create the phenotype, generally proceeds by methods that are universal across most phyla. Then is that transcription technique a part of the genotype specification when considering a new escalation of information content, or must the attribution only include those processes that are unique to the new clade or class? Also note that not all inherited DNA resides in the nuclei of the gametes that merge in sexual reproduction. We have already discussed the maternal contribution of mitochondrial DNA, and both gametes contain other cellular constituents as well. How, if at all, do we include these contributions?

On some levels answers to some of these questions can be given, but even then we must be sharply aware of implicit limits. Consider the example quoted by Quastler[9] — generating the organic molecules of a bacterium for their precursor constituents. We are on fairly safe ground if we recognize that these $10^{10}$ to $10^{13}$ bits reflect underlying chemical processes that have a universal nature. They can always be assumed to occur, under appropriate conditions, and need not be inscribed in the bacterial genetic code. The same can probably be said of the incorporation of those molecules into macromolecules, but their adaptation to make specific proteins or enzymes or to form cellular structure, is not as definite. This is part of the maximum of $10^5$ bits of bacterial genetic information.

As mentioned in Chapter 7, Gatlin[10] has dealt with part of this problem. She found, in vertebrate animals, that 70% of the genetic information is potentially meaningful code for the particular species and individual differences, though a major fraction of this is synonymous, i.e., not expressed. In additional, 70% of the rest, or about 20% of stored genetic information, is in the form of system "grammar," i.e., inherent interpretive procedure. It appears, therefore, that some "system strength" is included in the genetic information measure.

"Algorithmic" information of a pattern or structure[11] is an attempt to separate some of these aspects. This is defined as the length, in bits, of the minimum program that can be run on a generalized computer, so it can then duplicate the specific pattern or structure being evaluated. Since the computer is already in existence and operating as the mechanism of interpreting the algorithm, many items on Table 9-1 need not be included in Algorithmic information. This implies that at least some previous attainments are not to be included in considering the information change generating an escalation. It is not clear what other parts of the table need be considered since, e.g., the duplication of a physical structure is a different order requirement than is the replication of a pattern or plan. From the biological figures just quoted it appears that algorithmic information alone may not be sufficient, though its use is a helpful way of separating some contributions. Its relation to physical entropy, complexity, and other measures of information is actively under study. For example, Saunders and Ho[12] "define the complexity of a system to be the amount of information required to specify its construction." From this simple statement it seems to be the same as what we have called Taxonomic Complexity. However, their usage is equivalent to algorithmic information and their discussion, like that given above, and by others who propose complexity as the evolutionary parameter, deals with the same problem of inclusiveness.

To compound these problems of information definition we must recall the likelihood that the information variable sought has changed its nature during the cycles of evolution, as demanded by Price's condition for escalation. This implies that different aspects of information are likely to be significant during different evolutionary escalations. To repeat a statement from Chapter 8, during the cycle following Big Bang we have to deal with thermodynamic heat and local reduction of ordering entropy. This may have been followed by an escalation in which organic information storage ability developed, and then a cycle or cycles when physical complexity developed, when information utilization grew, and/or when information collection was significant. Perhaps there was a second interval of information processing ability, an escalation involving information generation, and there surely appears to have been one or several of information modification and information transmission.

Despite these differences and uncertainties the model used to derive the Evolutionary Trajectory implies one specific role that is played by previous informational level attainment. In Chapter 6 it is shown that at each escalation the logistic growth rate increases by a factor of $N/\eta_{j-1}$, the ratio of the evolving variable change to the once-removed, earlier saturation value. This has the effect of shortening the cycle time. It thus appears that the contributions of earlier escalations do not disappear when their own cycles are completed; they shift the relative growth base. For subsequent cycles the rate of change reflects the fact that it is starting from a greater, or more advanced, state.

## Limits on Entropy and Information

From the previous discussion we see that quantitative comparison of the magnitude of information between cycles is difficult, if not meaningless. Despite this, however, an important feature of the increase or, better, the change of information can be understood by considering the gross orders of magnitude of information-entropy change over different portion of the evolutionary trajectory.

In the period of astronomical change that, of all the evolutionary cycles, is closest to the pure physical origin of the entropy concept, it is unclear what specific system is to be considered when attempting to calculate entropy reduction. This is important because of the extremely large entropy increases that occur in the solar and galactic physical processes. Earlier we discussed the separation of massive particles and radiation shortly after the Big Bang. That involved both an ordering of these components, with its internal entropy reduction, and an accompanying greater overall increase of external entropy. It is estimated that much less than 1% of the entropy of our universe is associated with the massive particles constituting the physical portion of the universe, the rest being attributed to radiation.

In this regard we note that while the engineer measures entropy and information in bits, the physicist's entropy is measured in entropy units, abbreviated "eu." According to Equation 7-2 the units of eu are Joules per degree, derived from Boltzmann's constant, k. The conversion from one measure to the other[13] is only a matter of changing the logarithm base from 2 to 10 and inserting k.

$$\begin{aligned} S(\text{in eu}) &= 0.3010\, k' \times S(\text{in bits}) \\ &= 0.9570(10)^{-23} \times S(\text{in bits}) \qquad (9\text{-}1) \end{aligned}$$

Because of the smallness of $k$, the bit of information represents a very small part of an entropy unit. A 1000 page book in English, that transmits an average of 1.5 bits per letter, carries about $5(10)^{-13}$ eu. The entropy changes of physical systems are so much larger because of their very great number of particles that produce very high degeneracies. These entropy changes can also be found from the relation $\Delta S = \Delta Q/T$ so that, for example, to change only one gram of water from liquid to vapor at 100°C (373 K), that is, to boil it off, requires 2250 Joules so that $\Delta S$ = 2250 J/373 K ≈ 6 eu. It would take a pile of pages nearly a light-year high to contain as much information as the entropy change of this single gram of boiled water. This difference is striking and critical.

The large entropy change in physical processes is the ever increasing contribution considered in the Second Law of Thermodynamics. A very small fraction of the energy thereby liberated is taken by evolving systems to increase their information content and to decrease their own entropies. The two processes are in opposite directions and vastly different in magnitude, and should not be compared. While the former may be leading to a universal "heat death," the latter has led to local order and what we are.

## Driving Forces

While we have been primarily concerned with the timing of events along the trajectory, these considerations raise a different question, of just what are the mechanisms of change. Wide variation of the nature of the systems involved means that different forces drive each evolutionary cycle. However, the commonality of information, through its various aspects as a measure of the evolutionary trajectory, seems to imply that there is also an underlying unity to the diverse, detailed mechanisms.

During the initial, cosmic, cycle we can discuss thermal and mechanical forces that move the nonequilibrium, thermodynamic system toward some (perhaps unachievable) stable state. These involve reduction of free energies, phase changes, and relativistic dynamics, as well as other, more specific astrophysical mechanisms. As pointed out by Prigogine,[14] and as described earlier, these lead, naturally and inherently, to localized and general growth of information.

As for the intermediate cycles, the mechanism of biological change has been argued, in different forms, since well before Darwin's time, and is one of the most hotly debated questions of modern macroevolutionary discourse. The discussion at the beginning of Chapter 3 deals with this in some detail and, in particular, points out that the New Understanding of evolution suggests that genetic change may, at least in part, be driven by inherent, internal tendencies: the "Molecular Drive." To the extent that endogenous biological change is successful, we might expect that the tendency to experiment will also be enhanced, possibly resulting in an acceleration of the evolutionary pace. Since that is what the evolutionary trajectory appears to show, a word of caution is in order.

The most recent cycles contain the behavior and activities of intelligent creatures. There appears to be an "intellectual drive," in that the many aspects of information are rewarding to, or even necessary for, intelligence and have been sought after and developed.

It was explained in Chapter 6 that the logarithmic shortening of escalation cycles results from the multiplication of each successive growth rate by the inverse ratio of saturation levels. This reflects the higher level or population from which the next escalation begins. In his pioneering analysis of complexity and hierarchy, H. A. Simon[15] points out that the number of stable "subassemblies" increases the probability of restructuring and evolution in General System Theory. Our result is a quantitative realization of that fact. Thus, the question of endogenous enhancement is irrelevant to the observed shortening of the cycle times, although it may be one factor determining the magnitude of $r_0$.

The existence of the Evolutionary Trajectory causes us to rephrase a fundamental question posed by S.J. Gould and discussed in the introduction to this book. Rather than his formulation we ask whether the growth of information, implied by the Trajectory, is an indication of a directed process or is it merely an indication of the direction of progress? The first implies that evolution is controlled, calling for a teleological interpretation. While many regard this as a possible, and desirable, interpretation, it is supererogatory from all the available evidence. In the second view the normal mechanisms of change act in each cycle, based on local physical forces, or short term advantage. Here it is understood that the changes occur with no regard for the fact that the sums of the individual events result in movement and change over the long term. While the trajectory seems to imply a subtle law of behavior that underlies the changes we should examine that conclusion more carefully before accepting it.

The thermodynamic law of increasing entropy is a relevant example to consider in this regard. Given any particular physical situation, the immediate, local dynamics direct its change. Forces move bodies, heat flows from high temperature locations to those of lower temperature, etc. In each, the system changes to a more probable state, subject to the constraints of those dynamics and to the conditions of equilibrium. As we have seen, this implies greater degeneracy and therefore increased entropy. The entropy increase appears as a consequence of the other causes rather than being the driving force. Yet physicists and chemists can proceed as though entropy maximization can drive the system to its new state. One understanding of this lies in the law of large numbers, discussed earlier. When large numbers of particles are involved, the highest entropy state is so much more probable than any others that one can take it as a certainty. Maximum entropy increase occurs as surely as if it were the forcing condition. The one gram of water we considered earlier contains over $3 \times 10^{22}$ molecules, making its thermal behavior and its entropy change a certainty. We can state with confidence that its entropy will increase when it is boiled to steam, and by how much.

The numbers are very much smaller when dealing with information during evolution and, relatively, we are examining it with a much finer microscope. This is indicated by the magnitudes of the physicists' or chemists' entropy unit, Joules per degree, and the information theorists' unit, bits. Therefore, in any situation evolutionary fluctuation and deviations from maximum information could be *relatively* large. The situation with small numbers was illustrated by the three particle-three volume example in Chapter 6. That physical system spends 1/9 of the time in its lowest, rather than its highest entropy state. This is when all three particles are in a single subvolume.

This illustrates an important general proposition:

> *Processes that generate order are in no sense driven by the growth of entropy.*
>
> In particular, biological evolution is not driven by the growth of entropy.
>
> Layzer[4] — italics in original

In any particular case, the change of a species or societal development stage need not be in the direction to maximize, or even to increase its information. The resultant behavior gives the appearance that the macroevolutionary process is random and short-term. However, on the super-macro-evolutionary level of the trajectory the cumulative numbers seem to be large enough that a trend appears.

Therefore, one cannot conclude that evolution is *driven* by a law of increasing information or, to the extent that they are equivalent, by increasing complexity. The recognized driving mechanisms, such as survival-procreation and growth-expansion, remain paramount, and the basic questions of macroevolution are unchanged. On its various parts the trajectory must be interpreted to indicate that enhanced information is a result of cosmic dynamics, or that it promotes survival-procreativity and growth-expansion, or that it is rewarding to and sought after by intelligence. It can be used as a unifying *measure* of evolutionary success. It is difficult to make this relation more definite, and it is inappropriate to state it as a law since the statistics of small numbers leads us to expect that supporting examples can be found for any reasonable proposed mechanism and, also, that exceptions will nearly always exist. The Trajectory results from the fact that each successful stage acts as a springboard for escalation to the next stage. It presents a phenomenological description of the change of information with the advance of geologic time. We should also note the apparent urgency of informational change, as suggested by the fact that each successive escalation is initiated immediately upon completion of the previous one.

Our greater knowledge of the changing logistic variables of some of the other trajectories discussed in Chapter 6 may yield a better understanding of the mechanisms and reasons for the various escalations. Thus, the electronic and nuclear differences and stabilities of the chemical elements of Figure 6-1 clearly distinguish the constituents of each cycle. For the social changes shown in Figure 6-3 and 6-4 the underlying feature of that process is revealed by the extension of the religious trajectory into the domain of scientific understanding, which then, in turn, became engineering engagement with nature. This implies that it is the search for fundamental understanding of the world (and the concomitant desire to manipulate it) that is the underlying subject of concern. Each of the changes could only occur with understanding and within the evolutionary framework of its own time, but the trajectory reveals that a single search has never ceased. Even more, it implies that the collective mind of western society demands a "periodic" renewal or updating of its understanding. The old way, after its exciting introduction and growth, loses its vitality and become senescent, and conditions change, making it timely for a new escalation. This is a new concept for social change, that is demonstrated here.

# 10: CONSEQUENCES

## Overview

We have examined escalated logistic growth as a means of understanding the mechanism of evolutionary change. This suggested the construction of Figure 6-7 — the Evolutionary Trajectory — an empirical observation of the relation of a progression of system transition events. The relation exists on a super-macroevolutionary level and its coherency through successive escalations implies continuity of an underlying evolutionary parameter. Analysis of the process, particularly of its most recent cycles, establishes information as the evolving variable.

It should be emphasized that the Trajectory itself does not reveal why the transitions occur. In each case this is due to the immediate circumstances and dynamics of the individual situation. Some of these may be surmised from the developmental scenarios presented in Chapters 2, 3, and 4. The trajectory reveals only that they occur in the anticipated timely manner, with no delay beyond the uncertainties indicated on Figure 6-7.

Perhaps the most startling aspect of the Evolutionary Trajectory is its continuity, from the earliest cosmic moments, through the biologic life processes, through the occurrence of intelligence, to the products of that intelligence. This indicates that evolution, in the sense of local entropy reduction, began well before we normally conceive of it occurring, and continues without interruption, well beyond the point where it is usually regarded as ending. It extends to the present moment and, as we will show, into the future, although it has taken a heretofore unrecognized form.

It is impressive as well, to realize that it proceeds without deviation, at least on the scale with which we observe it, through events that should be expected to change its direction or alter its timing. What may have been the greatest extinction of animal life in earth's history occurred 250 million years BP in connection with a major glaciation, rapid atmospheric composition variation, a large sea level shift, and with the great Siberian lava flow and volcanic activity.[1,2] It has recently been suggested[3] that this may also be associated with a great asteroid impact with the earth that also initiated the breakup of the great, southern continent, Gondwanaland. This is not manifest from our progression of transition events. The same is true of the asteroid that struck the earth

65 million years ago marking the end of the geologic Cretaceous period and the demise of the dinosaurs and a great host of other creatures.[4] In this case the first mammals were already on the scene for five million years, and were able to attain prominence.[5] It is estimated that the earth has experienced five such impact induced mass extinctions in the last 600 million years.[6] And the ice ages that overcame the earth in the Pleistocene epoch seem, in this broad view, neither to have delayed nor encouraged the timely development of civilization. Closer to our own time we see that neither mankind's self-imposed wars and disruptions, nor the famines and epidemics imposed by nature, altered the timely development of writing, and then printing, and then electromagnetic communication and digital electronic computation.

We can conjecture whether these potentially "altering" events had more subtle effects. It is possible, even likely, that, following each one, subsequent evolution changed its direction even as it retained the cyclic timing we observe. Perhaps the world would be very different from what we now have if these events had not occurred. Our knowledge of the instantaneous dynamics of evolution indicates that it is a haphazard progression in which chance, opportunism, and competition determine each step. In one instant the most prolific group may thrive and become dominant, in another it may be that group with a small reproductive rate but which nurtures its progeny. The most suitable and fit in one circumstance may not be so in another.

Partly in response to this variability a speculative field of study, Exobiology, has appeared.[7] This deals with possible forms of extraterrestrial life and with evolutionary alternatives on earth. For example, had the dinosaurs not become extinct when they did, it is likely that our mammalian progenitors would have altered their evolutionary direction. They may have continued along the evolutionary trajectory, but with different results than we have today. Or they might have been suppressed by the dominant reptiles, allowing other avenues of evolutionary change. A recent study[8] has projected dominant reptilian evolution to the present, arriving at a modern counterpart of mankind.

This result reflects the belief, by some, that there are certain imperatives of evolution, i.e., that some developments are so natural and vital that they would be likely to occur regardless of the historical factors that control change. Thus the above study led to bipedality and to encephalization. In general, such studies are useful in that they lead to a critical analysis and weighting of the diverse factors involved in evolution. However, they probably represent only a fraction of the possible outcomes since nature has generally been more resourceful in its creativity than has the human mind. Even in the case of science fiction, where there are no limits other than imagination, F. Jacob[9] points out that

> The surprising point here again is what is considered possible. It is the idea, more than a hundred years after Darwin, that, if life occurs anywhere, it is bound to produce animals not too different from the terrestrial ones; and above all to evolve something like man.

Until we encounter the products of other evolutionary processes we cannot assess the uniqueness of what transpired on earth.

The quantitative nature of the trajectory introduces a new mode of analysis and new concepts into our understanding of evolutionary change. This is particularly striking if we extend it into the future. In the frontispiece quotation, Prof. Day notes that

> We apparently have reached a critical point in biological evolution. Either the trend of evolution is no longer valid or a radical change in the evolutionary process is imminent.

He is inaccurate only because the direction of change has already turned away from a biological path, with the works of mankind as its vehicle, and the "radical changes" are going on about us. The evidence of the Evolutionary Trajectory is that the logarithmic pace of evolution has not changed. As a result we are, indeed, facing "something momentous." Extrapolation of the Trajectory into the future shows that an evolutionary limit is fast approaching, and the time scale involved does indeed indicate "that the interval for man's preeminence [may] be ominously short."

Before treating this topic, however, we discuss the implications of an early concurrence in the evolutionary table, that bears on the very origin of life.

## The Immediacy of Life

On the tabulation of evolutionary transitions, in Chapter 6, event number one is stated as "Formation of the Earth and Initiation of Life." There are four events that should be distinguished within this description: condensation of the earth, formation of its solid crust, the appearance of replicating molecules, and the appearance of bacteria. Some

TABLE 10-1
RESOLUTION OF OCCURRENCES AT EVENT 1

| EVENT | TIME BP | log t |
|---|---|---|
| Formation of Earth | $4.6\ (10)^9$ | 9.66 |
| Appearance of Stable Crust | $3.6\ (10)^9$ | 9.56 |
| Replicating Molecules | undetermined | 9.54–? |
| Bacteria (microfossils) | $3.4\ (10)^9$ | 9.53 |

have recently assigned somewhat earlier dates to these events but this only slightly changes their relative times, listed on Table 10-1.

The separation of these points is small but discernible on the scale of the trajectory. Taking the midpoint of the first and last of them gives $4.0 \pm 0.6\ (10)^9$ BP for the median age of event 1. If we wish to consider the transition from astrophysical evolution to biological evolution it is probably more appropriate to take the middle two points, representing the change from cooling of the earth to the appearance of those complex molecules that are precursors of life. Although this first molecular appearance left no known fossil evidence, its age can be placed within the relatively narrow range allowed by the table.

As discussed earlier, it is currently acknowledged that simple organic molecules were extant on early earth and in our solar system debris. This somewhat shifts the emphasis of the critical change to life to the point at which a lipid envelope first surrounded the precursor replicating molecules. This allowed enrichment of the local molecular environment and constituted the earliest bacterial cell formation. On the appropriate, logarithmic scale the sought after difference of logt is about 0.02 to 0.03 logyr. This should be compared with the total cycle length between transition events, which has a logarithmic measure of about 0.67 logyr; the points are separated by less than 4% of a cycle. That the initiation of life occurred so precipitously after the appearance of conditions that tolerate it implies an immediacy of one to follow the other, or at least the lack of significant barriers. As Day points out:[10]

> So early did the primitive cells occur that the processes that led to their formation must have been highly probable. So probable, in fact, was the event of their appearance that it must be regarded as having been inevitable.

Although biologists and organic chemists agonize over understanding the initial moments of life and replicating molecule formation, nature seems to consider it trivial.

We have no other test of the hypothesis but we assume, from this inevitability, that it applies throughout the universe. Some form of life will evolve wherever conditions exist that can tolerate and support it. While many aspects of these conditions could be discussed, let us recognize that in one particular way the environment on the earth is cosmically rare. This refers to its abundance of heavy elements, which differs from the cosmic composition of roughly 90% hydrogen atoms and 9% helium atoms. Only these lightest elements were formed in the early stages of the universe, after the Big Bang. Carbon can form in the nuclear energy cycles of some hot stars but the generation of higher atomic number elements, such as nitrogen, oxygen, silicon, iron, etc., requires the extreme conditions found naturally only in Novas and, particularly, in Supernovas. Stars that are much more massive than the Sun expend their nuclear energy source of hydrogen at a prodigious rate. They pass from birth to senility in times of tens of millions of years, rather than the billions of years expected of sun-like stars. In its final stage such a star may explode as a Supernova and in the process heavy elements are created in its extremely high temperature, extremely high pressure interior.

With a Supernova explosion the heavy elements are spread into space. When new stars form from the enriched gas they, and any accompanying planets, are similarly enriched. The elemental composition of the earth indicates that our sun is such a second generation star.[11] Although this sequence of events is not common, because of the long time involved, and throughout the cosmos, the estimated number of daughter stars is tremendous. And so must be the number with planets whose size and solar distance would allow them to support life, and which must, therefore, bear life. The births and lives of first generation stars, and the subsequent births of second generation stars all involve random time periods. It is likely that some of these planets were formed and had life before earth. If their evolutions followed a similar trajectory to our own, then a range of parallel or divergent trajectory lines is possible. Later we discuss the fact that an initial small difference of these trajectories can become significant in the late stages of evolution so that the products of some of those extraterrestrial evolutions may well be at a more advanced stage than we.

## Future Transitions I

In constructing the Evolutionary Trajectory all times were denoted BP, i.e., Before the Present. We noted that taking the "Present" to be about 130 to 150 years from our reference of year 2000, i.e., by adding this period to all times, brought the most recent points onto the straight line of the earlier events. For convenience we take it to be 140 though this is not exact for several reasons:

- The ranges of the dates on Table A-1 lead to possible errors in the exact equation of the evolutionary trajectory.
- There is some uncertainty in the way the number 140 is derived, as explained in the Appendix.
- We should not necessarily expect transition points to fall exactly on the linear trajectory.
- As explained later, that date represents a change in fundamental nature of the trajectory and analogous situations frequently experience shifts in the occurrence of such changes.

However, the correction surely means that it is incorrect to take our reference to be the present rather than approximately, the year 2140 of the current calendar. Although the conclusions of the following discussion would still apply without this adjustment, the change allows us to assign specific times to potential future events. Extension of the trajectory line indicates that the next expected event, number 13, falls at 38 BP or on the date 2100. Further extrapolation gives, for point 14, the date 2130. These results beg for clarification.

Although evolutionary emergent events are abrupt, their suddenness is only relative to the time scale of observation. An observer who is aware of the immanence of such an event should be able to observe its development. For example, the anagenesis of biological evolution, i.e., its linear, steady change of species or types, has, by Darwin's own declaration, been considered to be too slow for human observation. However, at present, cleverly designed experiments are actually measuring some slow changes of incremental biological evolution.[12] Although a transition event represents a change that is unpredictable in form and function, it involves elements taken from its predecessor. One does not expect a precipitous departure from the previous forms. And the changes are outgrowths of processes that are already in place well before the transition. It is only after some development time that the true significance of the transition is visible. These behaviors must be particularly true for the present situation because the time interval is so small.

Therefore, in looking only 100 to 110 years into the future, to anticipate event 13, we should restrict ourselves to those scenarios that:

1. follow from the present condition and from currently visible trends,
2. are attainable within that scale of time interval,
3. involve information or are a consequence of it,
4. are technological in nature or have a major technological component.

And we must also be careful not to be swayed by imaginative, science fiction scenarios, though some such aspects are inevitable, looking as we are from the vantage point of a civilization, engaged in a scientific and engineering development mode. In performing this exercise we note that it has a considerable speculative element because the form in which a transition event appears is not as clear in advance as are some of the contributing factors.

With these guidelines and cautionary notes, we can probably eliminate a number of proposals that have been put forward.

• BIOLOGICAL EVOLUTION. Let us examine proposals that have been made for continued human evolution through (a) enhanced mental capabilities and (b) physical change.

(a) It has been suggested that human development is reaching a new, vital stage through the appearance of expanded mental capabilities that frequently go by the identifying name of psychic phenomena. Here we must avoid the frequent association of such capabilities as extrasensory perception with paranormal and supernatural influences. Through nearly a century of investigations of "parapsychology," noted and reputable physicists and psychologists have conducted experiments in various aspects of these phenomena. Some of their results appear quite convincing in having established "paranormal" behavior, i.e., an expansion of ability beyond what is considered normal based on accepted modes of behavior and laws of science. In studying these results one is, at best, dealing with extremely weak phenomena, or very low probability events, so that the difficulty is tremendous of conducting a truly proper, controlled test. Investigators are beset by systematic and random errors that are large compared with the phenomena being researched. Furthermore, the results are subject to statistical variations that allow even erroneous events to appear. Consequently, consistency has been characteristically lacking and even some instances of success do not establish the veracity of a phenomenon. Their investigators have been unable to convince the scientific community of the likelihood of these capabilities.

In this regard we note that, when discussing the left-right division of specializations in the brain, Norbert Wiener,[13] in his classic "Cybernetics," makes a statement that is, perhaps, to the point in considering the likelihood of further mental expansion. He comments that, for efficient functioning, "the closest possible association" is necessary between the hemispheres but that the existing interconnections are too few and are unreliable. He concludes with the statement

> That is, the human brain is probably too large already to use in an efficient manner all the facilities which seem to be anatomically present.... In man, the gain achieved by the increase in size and complication of the brain is partly nullified by the fact that less of the organ can be used effectively at one time.

(**b**) This is not to say that Darwinian evolutionary forces have ceased in our present civilization. Rather, as J. F. Crow[14] has pointed out, their influence and relative significance have been altered. Where we once faced and developed mechanisms for avoiding death from wild animals, our greater danger is now from disease, particularly communicable diseases. This is surely not a new phenomenon since our immune system has developed for just this reason, and, e.g., the sickling of hemoglobin cells is believed to protect against malaria. However our society is now confronted with diseases in which the immune system itself comes under attack and, where malaria is no longer a serious threat, sickle cell anemia represents a survival disadvantage. We have learned how to support these inherent protective systems by artificial means, such as immunizations, antibiotics, and chemicals. Whereas genetic diseases once eliminated susceptible individuals, we now have genetic counseling and tests to avoid some incidence and adversity, and the future promise is of genetic engineering to carry the process further. By these interventions we are altering the impact of factors that, in other times, acted as selective survival determinants.

Other factors that need be considered may not be so dramatic but, over the long range, and perhaps the no-so-long range as well, their influences may be of considerable importance as they act through our entire lives and over generations. Several examples can be given, such as the direct effects on our bodies and activities of industrial and automobile chemical pollution in the daily atmosphere. (The long-term, planetary effects are discussed later.) The pervasive use of food preservatives may be of great immediate health advantage but the truly long term, genetic effects are unknown. On a much simpler level, we are aware of the differences between the milk of different mammals, that have evolved to meet the needs of their own species' offspring. We may well inquire whether there is a subtle effect on human mental and physical development, of the widespread substitution, in our society, of modified cow's milk for mothers own, particularly during the critical period of infant brain growth.

While these examples indicate that we are altering natural selection, we should be aware that the overall quantitative effect is probably still quite small. This is indicated by two pieces of evidence. First, only a fraction of the world's population presently has significant access to, or is accessed by, the changes described above; for the majority, it is almost survival as usual. For the second evidence we refer to the prenatal mortality of domesticated animals, which is 30–40%. This is largely due to maternal-fetal disparities of blood type, or due to fetal genetic abnormality. If we assume that there is also an incidence of unrecognized miscarriages among women then, all together, we can conclude that the normal biological screening forces are still more-or-less unaltered.

Even apart from the considerations given in (a) and (b), above, the evolutionary record shows that biological changes evolve over periods of at least many thousands of years, major changes requiring hundreds of thousands of years. Although it has been close to 100,000 years since the last biological transition, the development of facile speech, we have seen that in the interim the mode of change has turned away from that avenue as the major evolutionary route. It is now directed to evolving man-made tools, i.e., artificial mechanisms with enhanced informational capabilities. While this does not proscribe eventual biological change, e.g., additional growth of mental capabilities or of immunological capability, it surely indicates that major natural change is unlikely within the next 110 to 150 years.

• THE WHOLE EARTH. It was pointed out in Chapter 3 that, through Natural Selection, the environment determines those creatures that survive, or at least selects for those characteristics with which it is conformable. On the other hand, the existing life, itself, largely *forms* the environment in which it exists and functions. On our finite planet all environments and life are ultimately interrelated. Atmospheric and water circulations provide common linkages between all species and all places, as do temperature changes, storms, etc. sea plankton and land plants control the water and atmospheric qualities that protects them and us from harmful ultraviolet solar radiation. Local and extensive food chains and life-death cycles, on land, sea, and air intimately relate all living things at all evolutionary and developmental stages. The thin layer that surrounds the inner earth, consisting of sea, soil, atmosphere and all the animal and vegetable phyla therein, constitutes a biological sphere within which members interact to maintain the whole. This is the "Whole Earth" concept.

James Lovelock and Lynn Margulis[15] have proposed that the Whole Earth is not merely a container of life-forms with their mutual effects but, rather, acts as though it is a living entity. It changes through time, varying so as to preserve itself, i.e., to retain the conditions suitable to the "biosphere" of which it is constituted. They have named this pseudo-lifeform "Gaia" after the ancient Greek goddess of the Earth. They point out that Gaia established itself on the naked earth and has successfully preserved itself through atmospheric changes, variations of solar intensity, and volcanic and meteoric explosions.

Early in its history Gaia was appropriated by those whose superficial understanding allowed them to impart a teleological purposefulness to earth's history and status, i.e., to associate supernatural powers to this new "life-form." More recently it has been acknowledged to be merely a convenient representation of the balance of natural influences. The concept of Gaia has been used to reinforce the belief that the recent activities of humanity, through air pollution, forest destruction, waste production, acid rain, over-fishing, land paving, warfare, etc., are occurring with sufficient intensity and rapidity that the biosphere cannot respond to maintain itself. To many it seems likely that, unless our disruption ceases, a new equilibrium will be established which may exclude much of life as we know it, with civilization and humanity not necessarily excepted.

The evolutionary trajectory has exhibited tremendous robustness through catastrophes of far greater magnitude than mankind's potential poisoning of the environment. While this gives us some confidence that it will continue, we have no guarantee of the nature of the subsequent conditions. In the processes we have been examining, on earth, there was always a ready substitute to replace a fallen leader in the evolutionary process. Since evolution is doubtlessly proceeding on multitudes of other planets in the galaxy it would be presumptive and unrealistic for us to assume that we are destined to lead in that greater competition. Nature cares little who advances and does not even pause over those who fail. Even if the trajectory of information continues we cannot ignore the possibility that, it may be in a world that is markedly changed and reduced from what we would like.

Many scientists, of widely different specialties, are studying this problem in parts and in its entirety. They have come to realize that to understand and control its complexity and vastness demand coordination on a scale that is far beyond today's grasp. The pertinence of the Whole Earth to our considerations would lie in the possibility that the next transition event might involve the acquiring of sufficient information processing ability and understanding that management of biosphere dynamics, becomes a possibility. These ideas and problems may be vital to <u>any</u> further evolution. However, as a goal in itself it has been noted by many that global environmental maintenance is more of a political, economic, and societal dilemma than it is technological. At least in the short term of decades it involves control and restriction of existing activities rather than a new transition stage of information storage and manipulation. If Whole Earth control is to be a vehicle for the development of a new order of computation and control intelligence then that aspect should be considered separately, and we do so in a later paragraph. As the evolutionary trajectory has put us in a mode of evolving man-made, technological means with enhanced information capabilities, Whole Earth management *per se* is unlikely to be transition 13.

• SOCIETAL CHANGES of other kinds can be ruled out for the same reasons as stated in the previous paragraph for the Whole Earth proposal. For example, Ralph Wendell Burhoe has put forward a Theory of Religious Biocultural Revolution.[16] He points out that the advent of civilization demanded large groupings of people to promote their individual and societal welfares, and these groups were initially formed based on religious beliefs. Today, competition between different religio-cultural entities has taken the role previously held by survival competition between species and families. Burhoe proposes that the next evolutionary stage requires resolution of this conflict.

Although religio-cultural groups have historically demonstrated the ability to merge, they are presently in a mode of confrontation. This is especially threatening because of the power of modern tools of warfare. Professor Burhoe's proposal seems to have particular currency in view of events at the close of the twentieth century which indicate that a major concern in the twenty-first may be global, ethnic and religious conflict. However, in terms of the evolutionary growth of information, Religious Biocultural conflict appears to be a side issue, much as the economic, political, and military conflict between Capitalism and Communism, that occupied most of the twentieth century. It did not lead to an evolutionary transition and it only tangentially effected development of the electronic digital computer.

• EXTRA-TERRESTRIAL CONTACT. Earlier in this chapter we indicated that there are surely other sites in the universe experiencing evolutionary dynamics.[17] Also recall our determination that some of them may have had beginnings before our own, because of the happenstance that their sun and planet condensed from some supernova detritus before ours. Perhaps this placed them on different evolutionary trajectories than ours, consistent with earlier attainment of evolutionary event number one. This is shown schematically on Figure 10-1. If their event "1" occurred $10^9$ years before ours (out of the roughly $12 \times 10^9$ elapsed years at that time) the location "elsewhere" would have now passed through three more transition events than earth. It might therefore be significantly more advanced. As with the historic intersection of more advanced Western civilization with less developed communities in Africa, North and South America, Pacific Islands, etc., contact with the civilization from "elsewhere" would markedly alter our own.

Having recently come to the full awareness that advanced intelligence must exist in our galaxy the scientific community has entered upon a program to search for signals sent by such intelligence.[18] Project SETI — Search for Extraterrestrial Intelligence — is an attempt to detect electromagnetic messages transmitted by a technologically advanced civilization that is trying to make its presence known. This involves scanning tens of

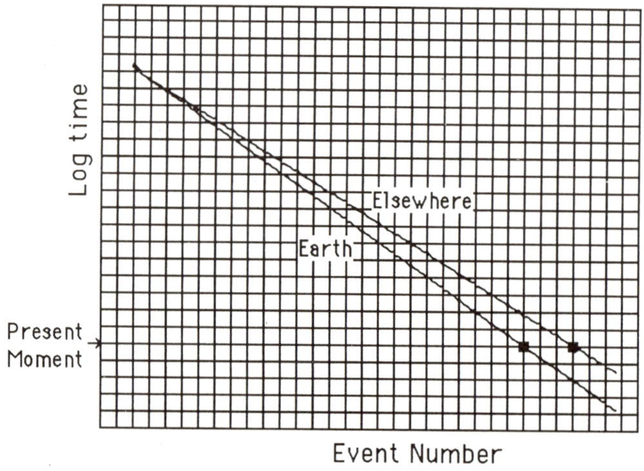


**Figure 10-1** Conjectured evolutionary trajectories at two different cosmic locations.

thousands of logical, natural, interstellar transmission channels with sensitive antenna detectors and sophisticated signal processing equipment to recognize any messages. Project SETI is partly based on the expectation that a more advanced civilization than ours will have developed the interest and capability to attempt purposeful intragalactic communication.

For an alternative scenario to the above we need only look to the trajectory of taxonomic complexity on earth (Figure 1-3). Its cyclic rate was greater than that of the informational trajectory and, had it not terminated after 9 escalations, it would now be 4 transitions more advanced. Therefore, although it is judging from only this single case, a weak speculation is that more rapid evolution may be subject to self limitation. More significantly, we should note that the evolutionary trajectory on earth begins at least as early as the Big Bang. Consequently we should not ignore the likelihood that our trajectory represents the universal rate of change, or nearly so. Of the only 1000 stars that are within 100 light years of our sun, the more advanced civilizations, described above and sought by SETI, are likely to be exceptions, though falling within the range of possible variations. Most other civilizations elsewhere will likely be at approximately the same developmental state as we. In that event the SETI search may be premature. This means that a negative SETI result, i.e., not detecting any extraterrestrial messages, will be as informative, but not as exciting, as a positive result. It could be taken to mean that our trajectory is fairly universal, at least within a cosmically local region.

In any event, even the advanced "elsewhere" appears not to be so much advanced that its communications will have sufficient intensity or have had time to traverse the vast interstellar distances to reach us. In this regard it has been pointed out that, from the early decades of this century, radio and television transmissions have made the Earth a source of electromagnetic radiation. This covers a fairly wide range of frequencies and might be expected to serve as an inadvertent indicator, to some distant technological beings, that other intelligent life has evolved in their (our) galaxy. Unfortunately these emanations are reduced by interstellar distance and are absorbed and scattered by interstellar dust and gasses. As a result we cannot expect our radiation to be noticed. For the same reason we are not likely to detect any alien transmissions unless they are being beamed directly toward earth by powerful transmitters.[19]

We note that gaining awareness of another technically accomplished galactic intelligence is very different from the tremendously more difficult task of communicating or, to an even greater degree, of achieving physical contact. Three or four more transition events presently seem unlikely to greatly change the immediate capability for interstellar travel, and we have not explored the possibility that there may be limitations on space travel.[14] These considerations, with the vast distances separating astronomical bodies and the short time since communication became possible, makes meaningful, evolution altering contact with extraterrestrials appear to be a poor expectation for event 13.

## Future Transitions II

The discussion in the previous section deals with scenarios that are unlikely candidates for the next evolutionary transition. Since we have seen that evolution describes an ongoing process there is every reason to expect that it will continue so we must still address appropriate possibilities for that event.

• INDUCED BIOLOGICAL CHANGE. We have concluded that natural biological evolution can be eliminated as a possibility for the near future. However certain advances of modern science, particularly in the past decade, have opened the possibility of "induced" biological change. Combining the refinements of modern chemical and physical techniques with the data handling, signal processing, and control capabilities of digital computers, has led to an understanding of the roles of, and the abilities to isolate and manipulate, specific genes and enzyme controls. New, heritable characteristics have been transferred from one species to another. The actions of genes can be enhanced, altered, delayed, or eliminated. Recent investigations have been opened into the control of senescence and cell death. These programs have been referred to, collectively, as molecular and genetic engineering.

Such efforts are dealing with modifications of the biological information storage and interpretation system. All sorts of marvelous or horrific prognoses can and have been made about the consequences of genetic meddling. At present, however, radical experiments are unacceptable by society and, in any event, most greatly altered life forms are not viable. Most effort in this field has been directed toward relatively minor modifications of plant and animal usefulness to humanity, and toward the understanding and control of specific human inherited disease and malfunction. At this time a scenario, such as of a new race, developed "in the laboratory" from *H. sapiens sapiens*, seems beyond exaggeration except, possibly, in the sense of a more disease resistant, healthier, more long-lived and productive population.

However, social and human values change in time and place and we cannot predict the direction of effort and change that will be initiated during the next 110 years by this exciting and promising research. It seems to be a rule that if something can be done it will be done. From the rate at which progress is being made, it is not unreasonable to foresee a transitional level change developing from this area of effort. Even though the evolutionary trajectory has turned away from natural biological change as its mode of transition, artificially induced changes in this direction could conceivably steer it back, and might well constitute an escalation of the type we are considering. Regardless of its transitional role, without doubt genetic engineering constitutes one of the important waves of the future.

• INTELLIGENT MACHINES. In Chapter 8 we briefly discussed efforts at computer simulation of the vast information storage and processing capability of the human brain. Our brain has orders of magnitude more neurons than there are elements in any computer, and has a vast complexity due to interconnections between individual neurons, networks of neurons, and between distant foci of activity and specialty. However, today vastly more memory and computational power can be contained on a single, centimeter square semiconductor chip than existed in the large laboratory full of electronics that constituted Eniac, the first large scale, digital computer, in 1944. Furthermore, the modern computer can act at speeds that are many orders of magnitude faster, and these capabilities are increasing exponentially with time. It may take a machine with the capability of processing $10^{15}$ bits, the estimated memory of the human brain, before such a thing as "true" artificial intelligence is possible.

Very recently proposals have been made, and preliminarily demonstrated for Quantum Mechanical and/or biochemical computers.[20] The earliest of these are attempts to overcome some of the ultimate limitations of present computer schemes by reducing component size to a level where individual electron, quantum mechanical, properties become primary. The advantages appear to be smaller size, greater speed, and reduced heat generation; the difficulties are reproducibility of component fabrication, accessibility to the components, statistical fluctuations, and the need for cryogenic cooling.

At a similar state of development are DNA computers, that use algorithmic parallels between problems to be solved and nucleotide structures. They utilize the vastly parallel configuration of large quantities of selected genetic material to produce the answer to a particular problem. This approach is too new for evaluation but it appears that its use may be limited to a small class of problems. If these schemes come to fruition their proponents estimate that they will result in speed increases and size reductions by several orders of magnitude, greatly facilitating the development of large computers.

In Chapter 5 we discussed the Principle of Self-organization. This postulates that large, complex systems and systems that are far from equilibrium may spontaneously enter a mode or modes of self-organization in which they create new patterns of behavior. It is not unreasonable to assume that this would apply to a PDP computer system with extensive neural network structure, and that is vastly more complex than exists today. Particularly if it has the indeterminacy associated with self-learning algorithms, such a development might possibly lead to a spontaneous new organization of the processing approach, i.e., to a new form of artificial intelligence. We cannot foresee what form this super-computer will take, or what functions it might serve, though some conjectures are presented below. However here again we must be wary of the science-fiction mentality. The hypothetical computer systems we are discussing must be vastly greater in complexity than any computing machines on today's horizon. Furthermore, present computers are severely restricted in response by their programming. While they may develop their own novel algorithms for handling the problems and data with which they deal, they are far from the "freewill" machines of science fiction imagination.

As with induced biological change, from the rate at which progress is being made, it appears that 110 years may be about the right period for developing such expanded computer capabilities. In any event, this area of research is expected to lead to new extensions of machine intelligence and information processing. Intelligent machines surely constitute another important wave of the future.

• MAN-MACHINE MERGER. In view of the recent evolutionary emphasis on artificial, i.e., man-made, modes of advancement, several authors[21] have proposed that an immanent development is a merging of man and machine. We are certainly aware of the common use of accessories, prosthetic devices, and transplants to improve, extend, and replace the function of human senses and body parts. Such methods are in their infancy and the promise of research seems endless. For example, new intelligent materials are being developed that respond to stimuli in ways that can be used as body aids. These include mechanical materials that approximate muscle, i.e., tension elements that remember their previous condition and return to it after a stimulus is removed. There are optical fibers that mimic or extend the sensitivity responses of the eye/optical system and considerable efforts are being made to duplicate the functions of the eye, including brain implants in visual sensory areas.[22] Attempts are underway to couple electrical signals between nerve and muscle systems and integrated circuits. Animal and artificial tissues are being developed to supplement or replace damaged or diseased human functions. If such efforts continue to expand and are successful one might envisage partly bionic people in the future.

The enhancements just described are of a physical-sensoral nature. In parallel with these we are witnessing an explosive enhancement of personal communication means and information access. This appears to be the initial consequence of system transition event number 12, the advent of electronic communication and digital computing. As the intimacy and pervasiveness of these features grow we can visualize greater expansion of individual capabilities to perceive, manipulate, and communicate information and to utilize computer intelligence, in real time, to interact with the environment.

Supported by conformable accessories and durable physical substitutes, and closely linked to powerful perceptional and information processing systems, the individual becomes a "cyborg" i.e., he or she is partly dependent upon, in the sense of closely allied with, artifacts in physical and mental functioning. Although we do not imagine the science-fiction threat of the machine portion dominating the human, this future population could have new and unique capabilities and abilities, and different perspectives and expectations. It might be considered to be the equivalent of a new species, thereby constituting an evolutionary transition.

Some of the scenarios presented in the previous paragraphs may seem fanciful. They are based on the suppositions of a tremendously rapid development rate through a period of over 100 years. It is common knowledge that developments in all the areas mentioned exhibit exponential growth, with doubling periods of between two and ten years, depending on the particular aspect considered. In view of this, 100 years is about the right period for the appearance of a system transition. This may be true as well for some of those we have eliminated, and others that have not been considered but that are surely extant, are varying for prominence. Like biological mutations that produce variations and create alternatives, they represent different technological techniques that produce variations and create alternatives. Biological variants respond to natural selection in an environment with parameters of predation, food, weather, etc. The alternatives being considered here must survive in an environment with parameters of political and societal suitability, and strongly subject to economic predation. Selection for survival is still an active process but we have long ago, at least since transition nine, left Darwinian selection and substituted the selection modes of human society.

## The Imminent Limit

We are presently passing through a period of unprecedented rapid change consequent to the evolutionary transition event of discovery of electromagnetic and electronic digital logic. The induced changes in the operating modes of society are radical and pervasive. We found, earlier, that extrapolation of the evolutionary trajectory to points 13, and 14, give the approximate dates 2100 and 2130. These are only 30 years apart — and further extrapolation leads to later event separations of about 4 years, and then even less. This raises the question of whether major changes can be tolerated at such small intervals without leading to chaos and a breakdown of the mechanisms by which they are absorbed into the fabric of being, growth, and society. Even today it is commonly observed that the pace, the rate of change, the inherent obsolescence of society and its tools and structures, is unparalleled, disruptive, and barely tolerable. How much more so when a series of major changes, each of which alters the very foundation and nature of the operation of society, occur in rapid succession. It appears that a limit of tolerable evolutionary progress is quickly approaching.

One possible result of this impeding crisis is the end of informational evolution. We have already considered two examples of similar terminations. For discovery of the elements, described in Chapter 6, there is a natural limit of physically stable nuclei. In evolution on earth the parameter is entropic, and we have no knowledge of its limits. The second example, that of taxonomic complexity, is perhaps, more revealing. That trajectory, of changing dominant life form, shown in Figure 1-3 ceased at event number XI. With the appearance of *H. s. sapiens* fundamental changes of the taxonomic aspect of evolutionary information has apparently ceased.

Even though one of its components has attained a limit, at present we have no reason to assume that the same behavior is inherent in the total informational evolutionary trajectory of Figure 6-7. As already mentioned, the informational variable is multidimensional and it is likely that one of its aspects can cease changing without altering progress of the others.

Then what might we expect beyond the year 2140? A hopeful alternative is that, once our evolutionary-informational maturity is attained, a new process of escalation will occur that starts a grand, new growth sequence. It is essential to make a distinction between the changes that have gone before and any such escalation we may anticipate after 2140. By that time the evolutionary trajectory presented in this book will have passed through approximately 14 major escalations during a period greater than $10^{10}$ years. As pointed out above, the entire super-macroevolutionary process will have progressed to a juncture beyond which it cannot continue, at least not in its usual mode. Any subsequent change would seem to call for a qualitatively different methodology, a "super-escalation" of a sort that is fundamentally distinct from its predecessors. We are looking to the beginning of an entirely new mode of change and a completely novel concept for the future; a conversion of the present evolutionary trajectory into something markedly different.

Within the context of normal logistics, De Solla Price discusses one possible response in a situation resembling that envisioned here. Consider that the process being examined changes its nature in such a way that the quantity being measured is no longer an appropriate description. The logistic curve must end and a new curve initiated, with a new variable as its measure and, probably, with a different rate of change. As an example, consider that milk production by farmers in some geographic region shows growth corresponding to a logistic curve. If market conditions change, so that these farmers find it advantageous to switch to beef production, the original measure, gallons per day, is incommensurate with the new one, pounds per season. The growth curve of milk production becomes meaningless and ends while the farmers continue to progress according to the new measure.

In the case of a change of "type," as this last scenario suggests, we are without concrete clues for guessing what will follow over the long term, or what will be its time scale. Doubtlessly the initial departure will be an outgrowth of existing developments. Because new growth curves initially change slowly, the time before significant change is apparent may not be short by our reckoning. However, one thing should be clear, even in this favorable scenario: Just as the Eucaryotes left behind the Procaryotes and the Hominids left behind all other life-forms, this new direction of change will eventually leave behind many of the values, systems, and devices of our glorious civilization. Professor Day's frontispiece statement "that the interval for man's preeminence will be ominously short" may be unduly severe and immediate in its threat, but it does seem to be an ultimate truth, at least in the sense of current expectations.

## Phase Change

Notwithstanding the uncertainties discussed above it is natural for us to grope for some possible understanding of the nature of change beyond event 13. Unfortunately, at present we can do little more than guess at its form. Although it is a method laden with hazards and pitfalls, in such cases it is often useful to examine an analogous, more familiar situation. An analogue is not equivalent to the particular case of interest, except in some limited sense that originates the analogy. One must not be led to accept it as representing that case more broadly, or to extrapolate its applicability beyond a limited range of correspondence. With this warming, and because we are otherwise at a loss for a way to proceed, we will resort to analogous reasoning.

Having drawn our understanding of entropy, system change, and information from Thermodynamics and Statistical Physics, let us turn back to that base for an analogue. We will develop a correspondence between physical phase change and the anticipated ultimate change in the nature of super-macroevolution. Two aspects of the analog will be considered; let us examine that relating to timing before that relating to mechanism.

TIMING: Figure 10-2 has two schematic curves that are derived from studies of ferromagnetic material. They illustrate the transition from magnetic to non-magnetic (more properly from the ferromagnetic to the paramagnetic) states. Physicists call this an "order-disorder" transition, or a second order phase change. Curve (a) shows the variation of the strength of magnetization, M, as the temperature is increased. The dashed portion is the extrapolated result of low temperature data. M should become zero as the ferromagnetic "Curie" temperature $T_F$ is approached. Instead it is found that thermal and statistical fluctuations, internal strains, and other local factors, cause M to persist until some higher temperature, the paramagnetic Curie temperature $T_P$, as given by the solid (a) curve.

If, on the other hand, we take the paramagnet and lower its temperature from some high value, we obtain a measure of its properties shown by curve (b). Here, downward extrapolation points to $T_P$ but, as this is

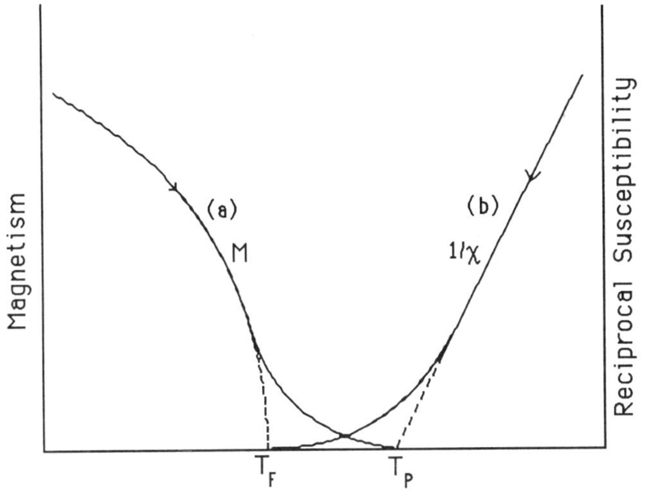


**Figure 10-2** Ferromagnetic phase transition.

approached the thermal and statistical fluctuations, internal strains, and other local factors delay the transition, approximately until $T_F$ is reached. Ordinarily these shifts are relatively small. In the ferromagnetic case $T_P - T_F$ is 10 to 30 degrees whereas $T_F$ is between 500K and 1000K, for different materials.

From these examples we wish to draw our first analog conclusion, that the time for the end of our present evolutionary history, i.e., 140 years hence, may well be subject to influences that cause a delay of its arrival. These would be factors that have ordinarily been relatively minor in timing of the evolutionary process. Very near the critical "phase change" the process becomes increasingly sensitive to them and alters its local behavior. In our present case these may be any of the normal, variable modes of societal behavior, such as a change of economic policy, a shift of scientific or national interest, national or institutional realignment, war or other social disruption, etc. The extent to which these factors can delay the terminus, due 130–150 years hence, its unknown but, from this analogue we do not expect the shift to be great.

MECHANISM: For a more detailed consideration of the analogue for change we begin by reviewing the degree to which the Ideal gas, considered earlier, actually represents a real gas. Recall that the equation of state for a fixed mass of gas, is

$$PV/T = \text{constant}$$

For an ideal gas this holds for all values of P, V, and T. As any of these change, the behavior described by this relation is smoothly varying. This description fits well with the behavior of most real gasses over their normal ranges of variation. However, our experience tells us that, e.g., if we lower the temperature of a real gas it eventually alters its nature in a radical way, i.e., it condenses to a liquid. For example, steam or water vapor become liquid water. Continued cooling produces a solid, i.e., ice. These transformations are called first order phase changes. As we know, they are reversible since heating melts the ice and subsequently changes the liquid water to steam. In our most common experience the phase transitions take place under conditions of constant pressure, that of the atmosphere in which our refrigerator and our tea kettle exist.

The schematic diagram in Figure 10-3a shows a typical form of the transformation boundaries between phases.[23] The horizontal, dotted line represents the situation just discussed, of changing temperature at constant pressure, $P_O$. The same sequence of changes can be induced in the laboratory by increasing or decreasing the pressure while the temperature is held constant. The vertical, dashed line represents such a path of changing pressure with the temperature held constant. We move down along this line by lowering the pressure along the isotherm, typically by pumping out the chamber holding the liquid. Eventually we arrive at a value $P_O$ on the phase dividing curve.

Crossing this curve causes the liquid to vaporize, for example, by boiling and evaporation. We have all had the experience of keeping a pot on a burner while its water boils. In so doing we add heat but the temperature (100°C or T = 373K) doesn't change. Instead the heat, $\Delta Q$, goes into the escaping steam. Since the thermodynamic definition of a change of entropy is $\Delta S = \Delta Q/T$, the system entropy is different before and after. The same thing occurs when the pressure is lowered. Crossing the phase transition curve of Figure 10-3a is accompanied by an entropy change. Figure 10-3b shows a different aspect of the phase change at $P_O$. The material entropy can be calculated from the laws of Thermodynamics and a linear relation results from plotting logP versus this entropy. The dotted line on Figure 10-3b shows that when we reach $P_O$, the pressure remains constant (along with the temperature, since we are are still in the isotherm) until enough heat is added that the entropy increases by the amount $\Delta S$. After reaching the vaporous state all the system variables can again change. Note that the inclinations of the lines

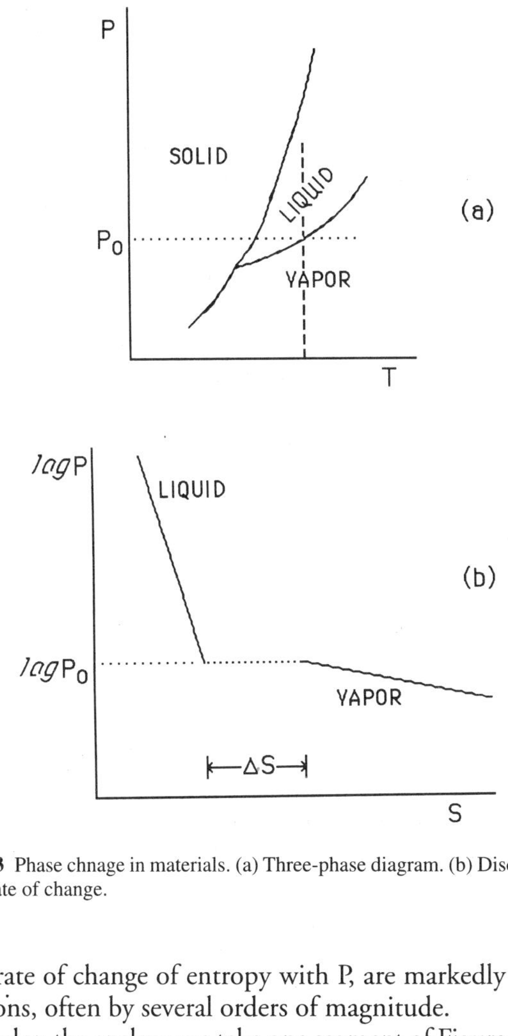


**Figure 10-3** Phase chnage in materials. (a) Three-phase diagram. (b) Discontinuity of entropy and of its rate of change.

i.e., the rate of change of entropy with P, are markedly different in the two regions, often by several orders of magnitude.

To develop the analogy we take one segment of Figure 10-3(b), say the liquid phase, to be the analogue of the present Evolutionary Trajectory, shown in Figure 6-7. The other segment is then the analogue of the trajectory after events 13 and 14. Since each cycle of Figure 6-7 represents an informational or, equivalently, an entropy change, the horizontal axis there corresponds to the horizontal axis here. LogP on Figure 10-3b is taken to be the analogue of logt.

Another feature of this physical phase change is important if we are to use the analogy to gain insight to the evolutionary change. The descriptions, above, and the dotted line of the discontinuity on Figure 10-3b indicates that the transformation from liquid to vapor involves a *process* during which S changes while P (and T) does not. Therefore the phase change is not as abrupt as it appears on Figure 10-3a. The transformation from liquid to vapor generally involves a considerable change of volume. For $P_O$ equal to normal atmospheric pressure a gram of liquid water at 100°C has a volume of just over one cubic centimeter. Its steam vapor at these conditions has a volume of 1617 $cm^3$. Thus, while pressure and temperature are constant during the phase transition, the process is accompanied by a very great change of density. On Figure 10-3 the density, or volume, is a hidden variable, but one whose alteration is an essential feature of the change.

To maintain the analogy we should seek a hidden evolutionary variable, whose discontinuous change is imminent at about year 2140. We must keep in mind that we are applying the analogy to the growth of informational ability. A guess as to the hidden variable can be found in a comment by Prof. Hans Moravec.[24]

> ... Our minds and genes share many common goals during life, but even then there is a tension between time and energy spent acquiring, developing, and spreading ideas, and effort expended towards biological reproduction ... The problem is partly one of time scales: humans already live extraordinarily long compared to other animals, no doubt to better teach their young, but the lifespan is a compromise with the genes' evolutionary imperative to experiment, the better to adapt ... all other things being equal, we'd likely be better off with a somewhat longer lifespan.

Professor Moravec then develops the case for redirection of the evolutionary pathway from human intelligence to intelligent robots. He reasons that the advantage of biological systems over machines has been gained by having evolved responsive perceptual and motor abilities. Intelligence follows from the world picture, i.e., the system of inherent experiences and responses, learned and displayed through these abilities. Therefore he postulates that the development of autonomous, learning, mobile robots, with greatly expanded computer memories, will eventually eliminate this advantage of organic lifeforms. Rudimentary examples of such robots are presently being constructed and tested.[25] In time, particularly with the development of biological and mechanical nanotechnology[26] (i.e., production of machinelike components and functional systems on a molecular size scale), vastly enhanced models should appear. With their longer lives for learning they can surpass their creators' informational abilities. It has been argued that their behaviors would be similar to, but superior to, those of organic life forms, and their creators may eventually be superfluous to their functioning and reproduction.

We should note that this proposed behavioral superiority seems to omit consideration of consciousness as a product of evolution. Self-awareness, intelligence, and appreciation of the experience of living (taking "appreciation" to imply *all* the varied connotations of that word) are properties of mind. Without becoming involved in a lengthy discussion of the definitions of these terms or of their realization, let us assume that they exist. They are usually considered to be connected with the brain, but to be additional to or beyond its mere physical functions. They appeared at some early point in evolutionary history and surely exist, to some extent, even in "lower" life forms. When a bat detects a moth, using its sonar emissions, it proceeds along an interception path that anticipates the moth's motion, that compensates for changes, and that avoids obstacles. By nearly every definition of the term the bat has an "image" of the world in which it functions and to which it responds. Since we do not even understand how our own visual image of the world becomes transformed into a coherent "comprehension" can we therefore doubt that some similar level of appreciation exists for the bat. In the same way, while we have difficulty understanding its existence in artificial life forms, at least in those of which we can conceive, it is not beyond imagination that some related property should come into being.

In this regard it has been suggested, and we mentioned earlier in this book, that behavioral fluctuations would be inevitable in any system approaching the complexity of the human brain, and even in much smaller systems. Particularly if it can adapt and accept changes these could generate unforeseen modes of external and internal analysis and operation. It is beyond our present understanding to know whether these could be equivalent, in some sense, to consciousness. In any event, Prof. Moravec believes that these robots would constitute the new wave of evolutionary advance.

He compares this development to the earliest days of the earth when, it has been proposed, clay minerals, particularly silicates, formed long chain, reproducing molecules. These served as templates for organic polypeptides which then replaced the clays to become the dominant genetic progenitors. This process has been referred to[27] as a "genetic takeover" and Moravec takes the "imminent" organic-to-machine transition to be a second takeover.

It was pointed out earlier that the machine "takeover" is but one possibility. Human life expectancy has increased more in the last century than in the last million years. And we are only beginning to understand its secrets. Chemical and genetic engineering and the microbiological revolution are presently solving the mysteries of some of humanity's most puzzling and serious maladies, and are even investigating the internal mechanisms of cell growth and death. Although some investigators believe there is an inherent limit to biological longevity it is, nonetheless, conceivable that human productivity and life expectancy will be extended to a degree we can barely imagine today. Of course we should not be surprised if phenotype changes accompany our genotype engineering. If these efforts are successful it may well be that we would barely recognize as our own, the long-lived, distant progeny that are produced and that constitute the new wave of evolutionary advance.

Note should be made of one other detail of Figure 10-3b. According to the analogue we have adopted, the change $\Delta S$ corresponds to a number of emergent logistic escalations, during which P, or its analogue, time, does not change. Being more realistic we can interpret this to imply a succession of very significant inventions and innovations over a relatively brief time. The exact meaning of "relatively brief" is not clear except that, in keeping with our earlier analogue of the delay caused by local influences, it may not be very brief according to human reckoning.

According to this analogy, then, an increase of longevity may be the hidden variable to produce an "evolutionary phase change." This might happen through human genetic alteration or through a "genetic takeover" by intelligent robots or by some combination of the two. The changeover can be expected to have precursors in 100 to 110 years and be enhanced about 140–150 years from now, and to initiate an indeterminate period of great change, followed by a different scale and kind of evolution.

## The End of History

Using the term "phase" in its more general sense, as a stage of development, in any case a major evolutionary phase change appears to be immanent. We mentioned earlier that reasoning by analogy is limited and dangerous. We have engaged in it to indicate one *conceivable* behavioral mode after our present evolutionary trajectory reaches its limit. We have pushed this analogy, probably more than is reasonable, to reach some rather detailed conclusions. Perhaps we have selected an inappropriate analogue for the discussion. Among material phase changes we have considered what is called a first order change. There are higher order phase changes that do not involve this kind of hidden variable. One such is the order-disorder change of the ferromagnetic analogy, that was otherwise not explored. Without doubt other analogs would lead to different expectations. One is easily led to "believe" an analogy and to follow its logic beyond the questionable parallelism that originally suggests it. Various investigators have followed the thermodynamic model to suggest, for example, besides the inverse correspondence of entropy and complexity, that Darwinian fitness corresponds to potential energy, and temperature to the degree of organization.

This impending evolutionary crisis has been referred to[28] as "the end of history." It is more appropriate to acknowledge that our distant intellectual or physical progeny have a future that we are not destined to share. The mortality of life-forms has always made this personally so, but now we may also sense a futility over our collective future. It is only in this sense that history will end. The political, social, artistic, scientific, and engineering activities of our minds should continue to maintain an active changing history. Our 10,000 years of civilization demonstrates that a great deal of activity, development, and change occurs without recourse to critical transitions.

The resort to analogy results from the opaqueness of future events, following the imminent limit of our present mode of evolutionary advance. We are loath to accept the possibility of that limit indicating an end of the path we have tread for ten billion years. Lacking any manifestation of the imminence of such an end we believe that evolution must continue. Since it appears that any future change will be at our instigation we should be cognizant of the responsibilities this incurs. As M. Eigen[29] points out:

> Now as always the motto of evolution is: survival. We will only be able to rise to this challenge by the mobilization of our mind, the ethical component of which is, however, unable to keep up with the breakneck pace of development in science and technology. Here, too, there is no 'world formula' to help us: step by step, we shall have to wrestle for the solutions ourselves, Man is still a relative newcomer to the planet Earth, and the creation of humanity has only begun.

— — — — — — — — — — — — — —

This book opened with a frontispiece quotation of William Day, anticipating the phase change we have explicated. It is, perhaps, appropriate to close with another from his same work.[10] In contemplating the immanent end of our evolutionary history he had already asked the question with which we are dealing:

> But what organizational level could possibly succeed us?

Although his response presupposes a particular scenario from among those we have considered, his perspective is revealing.

> Our biological bias binds us to our cosmic role. We regard life and consciousness as the ultimate forms of material realization and believe that man and his Darwinian successors will lead a conquest of space....
>
> ...But the universe does not share our biological bias. Life and consciousness are properties of biocreatures. In a cosmic evolution, biology is a phase that lies between chemical evolution and an evolution that is to follow....
>
> ...Man's technology represents the end of biological evolution's extent and the initiation of a mechanical evolution. Our devices are at the stage that chemical evolution was on primordial earth just before it created the first cells and ushered in the biological evolution. Machines are on the path toward self-sufficiency, like the biological cells. We may even build into them our consciousness and make them sufficiently independent that we inadvertently launch them on their own evolution.
>
> We may never know what we create, may never know the qualities that lie past consciousness and self-awareness. The advance into a larger and more evolved form will carry our creation into qualities beyond the pale of human comprehension-beyond our segment of the universe to a new dimension and a new reality.

For the true form of the future evolutionary trajectory we will have to await the revelation in time, be it through the hand and mind of man, the work of God, or the thought of Ptah.

# APPENDIX 1: DERIVATION OF THE EVOLUTIONARY TRAJECTORY

## Single Cycle Transition Time

Starting from the generalization of Equation 5-4

$$\frac{dx}{dt} = r_o x(1 - \frac{x}{X})^n \qquad \text{(A-1)}$$

and solving for $d^2x/dt^2 = 0$ gives, for the maximum values,

$$x\Big|_{max} = \frac{X}{1+n}, \qquad \frac{dx}{dt}\Big|_{max} = r_o X \frac{n^n}{(1+n)^{1+n}} \qquad \text{(A-2,3)}$$

A measure of the transition time from low x to X is

$$\Delta t = \frac{X}{\left.\frac{dx}{dt}\right|_{max}} = \frac{(1+n)^{1+n}}{r_o n^n} \qquad \text{(A-4)}$$

## Multicycle Transition Times

For the escalating system changing in the kth cycle, one can write equation 6-1 as

$$\frac{dx}{dt} = r_o x \prod_{j=1}^{k-1} (\frac{x}{X_j} - 1)^n (1 - \frac{x}{X_k})^n \prod_{k+1}^{max.} (1 - \frac{x}{X_j})^n \qquad \text{(A-5)}$$

This has a product of factors with $j < k$, the kth factor alone, and a product of factors with $j > k$. In the first product all terms have $x > X_j$ or, more likely, $x \gg X_j$ so the unity can be dropped; in the second product all terms have $x < X_j$ or, more likely, $x \ll X_j$ so these factors are close to unity. (A5) is, therefore,[A]

$$\frac{dx}{dt} = r_o \frac{x^{n(k-1)+1}}{\prod_1^k X_j^n} (X_k - x)^n \qquad \text{(A-6)}$$

From (A6) the maximum value of dx/dt is given at and by

$$x_{max} = \frac{n(k-1)+1}{nk+1} X_k, \tag{A-7a}$$

$$\left.\frac{dx}{dt}\right|_{max} = r_o n^n \frac{[n(k-1)+1]^{n(k-1)+1}}{(nk+1)^{nk+1}} \frac{X_k^{nk+1}}{\prod_{j=1}^{k} X_j^n} \tag{A-7b}$$

The transition time for interval k is

$$\Delta t_k = \frac{X_k - X_{k-1}}{\left.\frac{dx}{dt}\right|_{max}} \tag{A-8}$$

We can determine $\Delta t_{k+1}/\Delta t_k$, the transition times ratio of two successive escalations. This results is product of two terms:

$$\frac{\Delta t_{k+1}}{\Delta t_k} = \{*\}\left\{\frac{R_{k+1}-1}{R_k - 1} \bullet \frac{R_k}{R_{k+1}} \bullet R_{k+1}^{-nk}\right\} \tag{A-9a}$$

$$\{*\} = \frac{[n^2(k^2-1)+2nk+1]^{nk+1}}{(nk+1)^{2(nk+1)}} \bullet \frac{[n(k+1)+1]^n}{[n(k-1)+1]^n} \tag{A-9b}$$

where $R_k = X_k/X_{k-1}$.

The first curly bracket, a function of n and k, varies slowly (for n = 1 it changes by only 16% from k = 3 to k = 9) near the value unity; its slight variations can be ignored. The second depends on R; if R is constant, or changes slowly between cycles then the second curly bracket reduces to $R^{-nk}$. We therefore obtain a particularly simple dependence of elapsed times on cycle number, k. Defining $r = R^{-n} < 1$, we have

$$\Delta t_{k+1}/\Delta t_k \approx r^k \approx (X_{k-1}/X_k)^k \tag{A-10}$$

Taking the start of the first cycle as t = 0, the time to transition $X_k$ is

$$t_k = \sum_{j-1}^{k} \Delta t_j = \Delta t_1 \sum_{1}^{k} r^{k-1} = \Delta t_1 \frac{1-r^k}{1-r} = C(1-r^k) \tag{A-11}$$

where $C = \Delta t_1/(1-r)$ is not a function of k. Since $r_k$ is less than unity (A10) indicates that the series of escalation time converges. In fact, from (A11) we see that when k become large the limit is $t_\infty = C$, giving

$$t_\infty - t_k = C\, r^k \qquad \text{(A-12)}$$

For an escalation series that is completed we can shift the time datum to $t_\infty$ so that $t_\infty - t_k = t_{past}(k)$. Then

$$\log t_{past}(k) = -k\, n \log R + \log C \qquad \text{(A-13)}$$

This presents a linear relation between the logarithm of $t_{past}$ and cycle number. If $t_\infty$ is not known and a date is estimated which is too early by the amount $\tau$, then a MacLaurin expansion of (A12), leads to

$$\log t_{past}(k) = \log C - k\, n \log R - \tau/(C\, r^k) \qquad \text{(A-14)}$$

The points then fall increasingly below the linear relation as k increases. This derivation has assumed that n is constant. In the case that n changes slowly between cycles the derivation is more complicated but the forms of (A10) through (A14) are the same, except that n must be replaces by its average value for the cycles preceding the k–th.

## Endnote

[A] The approximations leading to Equation A-6 ignor the fact that the k – l and the k + l factors are very small near the lower and upper limits of the kth cycle, respectively. However, away from these ends, where one finds $x_{max}$, they can be used.

# APPENDIX 2: NUMERICAL ANALYSIS OF THE EVOLUTIONARY TRAJECTORY

## Regression Analysis

Regression analysis is a mathematical technique for obtaining the equation of a relation between two or more variables for which empirical values are known. In its simplest form it yields the parameters for a linear relation between paired data points. The most commonly used approach[1] is the method of least squares, as this yields equation parameters with the smallest variances, i.e., uncertainties.

In our case we denote the evolutionary event numbers from 0 to 10 by $x_i$. For each we take the time $t_i$ of the center of the range given on Table A-1. We can then calculate $y_i = \log t_i$. The set $(x_i, y_i)$ constitutes our data and

$$y(x) = mx + b \tag{A-15}$$

is the relation characterizing the evolutionary trajectory; we seek values of m and b. For each point $(x_i, y_i)$ the technique is to take the deviation

SUMMARY TABLE A-1
EVENTS OF THE EVOLUTIONARY TRAJECTORY

| NUMBER | EVENT | DATE (BP) | |
|---|---|---|---|
| 0 | Big Bang | 1.3–2.0 | $(10)^{10}$ |
| 1 | Solidification of Earth and Initiation of Life | 3.4–3.6 | $(10)^{9}$ |
| 2 | Eucaryotic Radiation | 5–7 | $(10)^{8}$ |
| 3 | Mammal | 1.5–2 | $(10)^{8}$ |
| 4 | Hominoid | 2.5–4.0 | $(10)^{7}$ |
| 5 | Hominid | 4–10 | $(10)^{6}$ |
| 6 | Homo | 1.5–2 | $(10)^{6}$ |
| 7 | H. Sapiens | 2–5 | $(10)^{5}$ |
| 8 | H.S. Sapiens | 7–10 | $(10)^{4}$ |
| 9 | Civilization | 1–2 | $(10)^{4}$ |
| 10 | Writing | 3–5 | $(10)^{3}$ |
| 11 | Printing | 545* | |
| 12 | Digital Computing & E-M Communication | 55* | |

* Times with respect to the calendar year 2000

$$e_i = y_i - y(x_i) = y_i - (mx_i + b) \qquad \text{(A-16)}$$

and to determine the values of *m* and *b* that minimize $\Sigma e_i^2$, the sum of the squares of the deviations. Since the squares are all positive this process guarantees an absolute "best fit." The sum is over the first eleven paired points; we have i from 0 to 10. The references on statistical analysis[1] show that after calculating the data averages

$$\langle x\rangle = (\Sigma x_i)/n \ , \quad \langle y\rangle = (\Sigma y_i)/n \qquad \text{(A-17)}$$

where the number of points n = 11, one takes the deviations $\Delta x_i = x_i - \langle x\rangle$ and calculates the parameters from

$$m = \Sigma(\Delta x_i)y_i/\Sigma(\Delta x_i)^2 \quad \text{and} \quad b = \langle y\rangle - m\langle x\rangle \qquad \text{(A-18)}$$

In substituting values in these equations we should note that the $t_i$ values, and therefore the $y_i$, are known to two significant figures only. For purposes of numerical manipulations intermediate results can be written to three of four places but the final results should be rounded to the same two figures, or at most three. Applying equations (A-18) to the data from Table A-1 yields

$$m = -0.6647 \pm 0.0045 \qquad \text{(A-18a)}$$
$$b = 10.214 \pm 0.030 \qquad \text{(A-18b)}$$

Letting i = k we can therefore write (A-15) as

$$\log t = 10.2 - (2/3)k \qquad \text{(A-19)}$$

## Extrapolation Points

For those points that were not used in the evaluations of m and b we can use Equation (A-19) to extend the trajectory to their locations. Table A-2 shows the anticipated values. Events 11 and 12 are of great interest. They occurred 545 and 55 years before the year 2000, which would give y values

TABLE A-2
OCCURRENCE OF EXTENSION POINTS

| EVENT NO. ($x_i$) | CALCULATED VALUE ($y_i$) | DATE ($10^{y_i}$) BP |
|---|---|---|
| −1 | 10.878 | $7.56\ (10)^{10}$ |
| 11 | 2.902 | 793 |
| 12 | 2.238 | 173 |
| 13 | 1.573 | 37 |
| 14 | 0.908 | 8 |

of 2.736 and 1.740, respectively. These are less than the calculated values on Table A-2 and they can be brought to better agreement by adding a time "T" to the actual ages before 2000. This implies that the reference date for the notation BP should be 2000 + T.

Consistent with the method of least squares we can obtain T by minimizing the sum of the squared deviations from their calculated trajectory values. For the two points involved we have

$$\Sigma e_i^2 = [y_{11} - \log(545 + T)]^2 + [y_{12} - \log(55 + T)]^2 \qquad \text{(A-20a)}$$

which yields the value T = 126 years. An alternative method is to have the event points lie equidistant from and on opposite sides of the trajectory. This is equivalent to requiring that

$$y_{11} - \log(545 + T) = -[y_{12} - \log(55 + T)] \qquad \text{(A-20b)}$$

which yields the value T = 145 years. These values are close enough that we need not be concerned with their difference. We will assume the value T = 135 years, with the understanding that this has only decade accuracy at best and, since points can be expected to deviate from the exact trajectory, even this should be taken loosely. Then the reference for our BP scale is the year 2135. When this value of T is added to the ages of events 11 and 12, before 2000, we arrive at the corrected ages, BP, shown on Table A-3 and on Figure 6-7.

TABLE A-3
CORRECTION TO TABLE A-1 EVENTS OF THE EVOLUTIONARY TRAJECTORY

| NUMBER | EVENT | DATE (BP) |
|---|---|---|
| 11 | Printing | 680* |
| 12 | Digital Computing & E-M Communication | 190* |

* Dates adjusted to reference year 2135

Finally, we note that the anticipated occurrence of events 13 and 14 are at the calendar years.

event 13: $2135 - 37 = 2098 \sim 2100$
event 14: $2135 - 8 = 2127 \sim 2125$

We refer to these dates in Chapters 9 and 10.

# REFERENCES


## Introduction

1. Norbert Wiener, *Cybernetics or Control and Communication in the Animal and the Machine* (MIT Press, 1961).
2. V. F. Turchin, *The Phenomenon of Science* (Columbia Univ. Press, 1977).
3. A. Jdanko, "On an Evolutionary and Negentropic Interpretation of Cybernetics," *Cybernetica*, 34 (1991): 173–175.
4. S. J. Gould, "Eternal Metaphors of Palaeontology," in *Patterns of Evolution*, edited by A. Hallam (Elsevier, 1977): 1–26.

## Chapter 1

1. W. C. Dampier, *A History of Science* (Cambridge Univ. Press, 1949).
2. E. McMullin, *Evolution and Creation* (Univ. of Notre Dame Press, 1985).
3. W. L. Stokes, *An Introduction to Historical Geology* (Prentice Hall, 1960); E. L. Simon, *Primate Evolution* (Macmillan, 1960).
4. M. M. Schneersohn, "A letter on Science and Judaism," in *Challenge*, edited by A. Carmell and C. Domb (Feldheim, 1988): 142–149.
5. G. Emiliani, "Stratigraphic Correlation and the Pleistocene Epoch," in *Time and Stratigraphy in the Evolution of Man*" (The Division of Earth Sciences of NAS, NRC, 1967): 56–62; W. T. Peckra and M. Rubin, "Absolute Dating and The History of Man," *Ibid.*: 43–56.
6. A. Moles, *Information Theory and Esthetic Perception* (Univ. of Illinois Press, 1966).
7. W. Gough, J. P. G. Richards and R. P. Williams, *Vibrations and Waves*, (J. Wiley & Sons, 1983).
8. J. W. Schopf, *Major Events in The History of Life* (Jones and Bartlett, 1992).
9. R. L. Cann, M. Stoneking and A. C. Wilson, "Mitochondrial DNA and Human Evolution," *Nature*, 325 (1987): 35–36; L. Vigilant et al., "African Population and the Evolution of Human Mitochondrial DNA," *Science*, 253 (1991): 1503–1507; Technical Comments, "Human Origins and Analysis of Mitochondrial DNA Sequences," *Science*, 255 (1992): 737–739.
10. L. L. Cavalli-Sforza et al., "Reconstruction of Human Evolution," in *Proc. Nat'l. Acad. Science, USA*, 85 (1988): 6002–6006.
11. F. J. Ayala, A. Escalante, O'hUigin and J. Klein, "Molecular Genetics of Speciation and Human Origins," in *Tempo and Mode in Evolution*, edited by M. Fitch and F. J. Ayala (National Academy Press, 1995): 187–211.

## Chapter 2

1. J. B. Prichard, *The Ancient Near East,* Vol. 1 (Princeton Univ. Press, 1958).
2. D. Corish, "The Beginning of the Beginning in Western Thought," in *The Study of Time, IV,* edited by T. Fraser, N. Lawrence and D. Park (Springer-Verlag, 1981): 34–45.
3. D. H. Weinberg, "Origins," *Science,* 320 (1985): 15–18.
4. A. Linde, "Particle Physics and Inflationary Cosmology," *Physics Today* (9/1/87): 61–68.
5. M. Bartusiak, "Before The Big Bang, The Big Foam," *Discovery Magazine* (Sept. 1987): 76–83; M. Waldrop, "Bubbles Upon the River of Time," *Science,* 215 (1982): 1082–1083.
6. M. J. Waldrop, "COBE Confronts Cosmic Conundrums," *Science,* 247 (1990): 411–413.
7. R. B. Partridge, "The Seeds of Cosmic Structure," *Science,* 257 (1992): 178–179.
8. S. van den Bergh, "The Age and Size of The Universe," *Science,* 258 (1992): 421–424; M. Rowan-Robinson, *The Cosmological Distance Ladder* (Freeman Press, 1985); J. Travis, "Hubble War Moves to High Ground," *Science,* 266 (1944): 539–541.
9. M. Waldrop, "Pushing Back the Redshift Limit," *Science,* 239 (1988): 727–728.
10. H. E. Newsom and K. W. W. Simms, "Core Formation During Early Acretion of the Earth," *Science,* 252 (1992): 926–933.
11. W. Day, *Genesis on Planet Earth* (Yale Univ. Press, 1984).
12. C. E. Chyba et al., "Cometary Delivery of Organic Molecules to Early Earth," *Science,* 249 (1990): 366–373; Also see ref. 10.
13. S. L. Miller, "The Prebiotic Synthesis of Organic Compounds," in *Major Events in the History of Life,* J. W. Schopf, editor (Jones and Bartlett, 1992): 1–28.
14. S. G. Cairns-Smith, "Beginnings of Organic Evolution," in *The Study of Time, IV,* edited by J. T. Fraser, N. Lawrence and D. Park (Springer-Verlag, 1981): 15–33.
15. M. J. Waldrop, "Did Life Really Start in an RNA World," *Science,* 246 (1989): 1248–1249.
16. W. H. Li and D. Grauer, *Fundamentals of Molecular Evolution* (Sinour, 1991).
17. L. Margulis, *Origin of Eucaryotic Cells* (Yale Univ. Press, 1970).
18. M. R. Walter, "The Timing of Major Evolutionary Innovations," in *Rates of Evolution,* edited by K. S. W. Campbell and M. F. Day (Allen and Unwin, 1987): 15–38; W. F. Doolittle & J. R. Brown, "Tempo, Mode, the Progenote, and the Universal Root," in W. M. Fitch and F. J. Ayala, eds., Tempo and Mode in Evolution, Nat'l. Acad. Sci. Press (1995): 3–24.

19. S. A. Bowring et al., "Calibrating Rates of Early Cambrian Evolution," *Science*, 261 (1993): 1293–1298.
20. L. Margulis and D. Sagan, *Origins of Sex* (Yale Univ. Press, 1986).
21. S. J. Gould, *Eight Little Piggies* (Penguin Books, 1993).

## Chapter 3

1. D. E. Koshland, "The DNA Repair Mechanism," *Science*, 266 (1994): 1925.
2. W-S. Li and D. Graur, *Molecular Evolution* (Sinauer Assoc., 1991).
3. Steven J. Gould," Speciation and Sorting as the Source of Evolutionary Trends," in *Evolutionary Trends*, edited by K. J. McNamara (Univ. of Ariz. Press, 1990: 3–27; — "Opus 200," in *Natural History* (August 1991): 12–18.
4. A. Gibbons and V. Morrell, "On the Many Origins of Species," *Science*, 273 (1996): 1496–1502.
5. D. J. DePew and B. H. Weber, "Innovation and Tradition in Evolutionary Theory," in *Evolution at a Crossroads: The New Biology and the New Philosophy of Science*, edited by D. J. DePew and B. H. Weber (MIT Press, 1985): 227–260.
6. J. H. Campbell, "An Organizations Interpretation of Evolution," *ibid.*, 133–167; — "The New Gene and Its Evolution," in *Rates of Evolution*, edited by K. S. W. Campbell and M. F. Day (Allen and Unwin, 1987): 283–309.
7. R. E. Lenski and J. E. Mittler, "The Directed Mutation Controversy and Neo-Darwinism," *Science*, 259 (1993): 188–194.
8. D. S. Thaler, "The Evolution of Genetic Intelligence," *Science*, 264 (1994): 224–225; J. Travis, "Possible Evolutionary Role Explored by 'Jumping Genes'," *Science*, 257 (1992): 884–885; C. J. Farr and P. N. Goodfellow, "Hidden Messages in Genetic Maps," *Science*, 258 (1992): 49.
9. F. J. Ayala, "Reduction in Biology — A Recent Challenge," in D. J. DePew and B. H. Weber, *op cit.*: 65–79.
10. E. Mayr, "How Biology Differs from the Physical Sciences," in *Evolution at a Crossroads: The New Biology and the New Philosophy of Science*, edited by D. J. DePew and B. H. Weber (MIT Press, 1985): 43–63.
11. B. Runnegar, "Evolution in The Earliest Animals," in *Major Events in The History of Life*, edited by J. W. Schopf (Jones and Bartlet, 1992): 65–93.
12. P. A. Moody, *Introduction to Evolution* (Harper & Row, 1953); A. S. Romer, "Phylogeny and Behavior with Special Reference to Vertebrate Evolution," in *Behavior and Evolution*, edited by A. Roe and G. G. Simpson (Yale Univ. Press, 1958): 48–75.
13. Robert T. Orr, *Vertebrate Biology* (W. B. Saunders, 1961); E. H. Colbert, *Evolution of the Vertebrates* (Wiley, 1955).

14. A. B. Smith and C. Patterson, "The Influence of Taxonomic Method and The Perception of Patterns of Evolution," in *Evolutionary Biology*, Vol. 23, edited by M. Hecht & B. Wallace (Plenum Press, 1988): 127–216.
15. V. Morell, "Life's Last Domain," *Ibid.*: 1043–1045; C. J. Bult, "Complete Genome Sequence of the Methanogenetic Archeaeon, *Methanococcus jannaschii*," *Science*, 273 (1996): 1058–1073.
16. W. L. Stokes, *An Introduction to Historical Geology* (Prentice Hall, 1960).
17. R. F. Foster, *General Geology* (Merrill, 1969).
18. R. Lewin, "How Did Humans Evolve Big Brains?," *Science*, 216 (1982): 840–841.
19. E. L. Simon, *Primate Evolution* (Macmillan, 1972).
20. W. E. L. Clark, *History of the Primates* (Univ. Chicago Press, 1959).
21. G. Brauer, "The Evolution of Modern Humans," in *The Human Revolution*, edited by P. Mellars and C. Stringer (Edinburgh Univ. Press, 1989): 123–154; P. Mellars and C. Stringer, "The Human Revolution," *Ibid.*: 1–13.
22. E. L. Simons, "Human Origins," *Science*, 245 (1989): 1343–1350; S. M. Stanley, *The New Evolutionary Timetable* (Basic Books, 1981); C. B. Stringer and P. Andrews, "Genetic Fossil Evidence for the Origin of Modern Humans," S*cience*, 239 (1989): 1362–1368; M. Titiev, *The Science of Man* (Holt, 1954).
23. P. H. Harvey and J. R. Krebs, "Comparing Brains," *Science*, 249 (1990): 140–146.
24. H. Jerison, *The Evolution of the Brain and Intelligence* (Academic Press, 1973).
25. S. J. Gould, *Ontogeny and Phylogeny* (Harvard Univ. Press, 1977).
26. A. H. Schultz, *The Life of Primates* (Universe Books, 1969); S. Ackerman, "European Prehistory Gets Even Older," *Science*, 246 (1989): 28–30.
27. C. B. Stringer and P. Andrews, o*p cit.*, ref. 24; R. Lewin, "Modern Human Origins under Close Scrutiny," *Science*, 239 (1988): 1240–1241; R. Lewin, "Species Question in Modern Human Origin," S*cience*, 243 (1989): 1666–1667.
28. T. Dobzhansky, "On the Evolutionary Uniqueness of Man," in *Evolutionary Biology*, Vol. 6, edited by T. Dobzhansky, M. K. Hecht and W. C. Sterre (Appleton-Century Crofts, 1972): 415–430.
29. A. Gibbons, "Neanderthal Language Debate," *Science*, 256 (1992): 33.
30. P. Lieberman, "Primitive Vocalizations and Human Linguistic Ability," *J. Acoustic Soc. Amer.*, 44(16) (1968): 1574–1584; — "The Origin of Some Aspects of Human Language and Cognition," in *The Human Revolution*, edited by P. O. Mellore & C. D. Stryer (Edinburgh Univ. Press, 1984): 391–414; — *The Biology and Evolution of Language* (Harvard Univ. Press, 1984); — "On Human Speech, Syntax, and Language," *Human Evolution*, 3 (1988): 3–18.
31. E. S. Crelin, *The Human Vocal Tract* (Vantage Press, 1987).

32. Norman Geschwind, *The Brain* (Freeman & Co., 1979); Jagjit Singh, *Information Theory, Language, and Cybernetics* (Dover, 1966).
33. J. L. Bradshaw and L. J. Rogers, *The Evolution of Lateral Asymmetries, Language, Tool Use, and Intellect* (Academic Press, 1993).
34. William H. Calvin, *The Throwing Madonna* (McGraw-Hill, 1983).
35. R. Lewin, "The Origin of the Modern Human Mind," *Science*, 236 (1987): 668–670.
36. L. C. Eiseley, "Man and Novelty in Time and Stratigraphy," in *The Evolution of Man* (Symposium Publication 1469, NAS., NRC, 1967): 65–81.

## Chapter 4

1. G. Emiliani, "Stratigraphic Correlation and the Pleistocene Epoch," in *Time and Stratigraphy in the Evolution of Man* (Div. of Earth Sciences of NAS, NRC, 1967): 56–62; W. T. Peckra and M. Rubin, "Absolute Dating and The History of Man," *ibid.*: 43–56; W. E. LeGros Clark, *History of Primates* (Univ. Chicago Press, 1959).
2. R. A. Foley and P. C. Lee, "Finite Social Space, Evolutionary Pathways, and Reconstructing Hominid Behavior," *Science*, 243 (1989): 901–907; R. N. Brandon, "Adaptation Explanations," in *Evolution at a Crossroads*, edited by D. J. DePew and B. H. Weber (MIT Press, 1985): 81–96.
3. R. G. Hepper, *Kin Recognition*, (Cambridge Univ. Press, 1991); R. Lewin, "Practice Catches Theory in Kin Recognition," *Science*, 223 (1984): 1049–1051.
4. C. D. Darlington, *The Evolution of Man and Society* (Simon & Schuster, 1969); D. H. Trump, *The Prehistory of the Mediterranean* (Yale Univ. Press, 1980); T. W. Walbank, A. M. Taylor and N. M. Bailkey, *Civilization-Past and Present* (Scott, Foresman & Co., 1962).
5. A. H. Paterson et al., "Convergent Domestication of Cereal Crops," *Science*, 269 (1995): 1714–1718.
6. C. W. Cowan and P. J. Watson, *The Origins of Agriculture* (Smithsonian Inst. Press, 1992); J. May, *The Progress of Man* (Pergamon Press, 1966).
7. E. K. Coughllin, "Farming Lessons From Prehistory," *Chronicle of Higher Education,* 17 February 1995, p. A10–A12.
8. M. Maxwell, *Human Evolution* (Columbia Univ. Press, 1984).
9. W. H. Calvin, *How the Shahman Stole the Moon* (Bantam Books, 1991).
10. I. J. Gelb, *Study of Writing* (Univ. Chicago Press, 1952).
11. D. Shamandt-Bessert, *Before Writing* (Univ. Texas Press, 1992).
12. D. C. McMurtrie, *The Book* (Oxford Univ. Press, 1943).
13. D. Ribiero, *The Civilizational Process* (Smithsonian Press, 1968).
14. R. Reif, "Browsing Through The Art of Gutenberg's Galaxy," *N. Y. Times*, 6 February 1994, p. 32.

15. K. Haefner, "Evolution of Information Processing-Basic Concept," in *Evolution of Information Processing Systems*, edited by K. Haefner (Springer-Verlag, 1992): 1–46.
16. J. D. Ryder and D. G. Fink, *Engineers and Electrons* (IEEE Press, 1984).
17. G. Himmelfarb, "Revolution in the Library," *American Scholar*, June 1997.

## Chapter 5

1. J. W. Kane and M. M. Sternheim, *Physics* (Wiley, 1988): sect. 8.6.
2. M. M. Schiffer and L. Bowden, *Mathematics in Science* (Math. Ass'n. of America, 1984); J. M. Smith, *Mathematical Ideas in Biology* (Cambridge Univ. Press, 1968).
3. D. J. De Sola Price, *Little Science, Big Science* (Columbia Univ. Press, 1963).
4. R. H. MacArthur and E. O. Wilson, *The Theory of Island Biogeography* (Princeton Univ. Press, 1967); J. J. Sepkoski Jr., "A Kinetic Model of Phanerozoic Taxonomic Diversity," *Paleobiology*, 4(3) (1978): 223–251; R. Pool, "Is it Chaos, Or Is It Just Noise", *Science*, 243 (1989): 25–28.
5. D. Gulick, *Encounters with Chaos* (McGraw-Hill, 1992).
6. A. I. Oparin, "The Appearance of Life in The Universe," in *Exobiology*, edited by C. Ponnamperuma (N. Holland, 1972): 1–15.
7. G. J. Vermeij, *Evolution and Escalation* (Princeton Univ. Press, 1987).
8. V. F. Turchin, *The Phenomenon of Science* (Columbia Univ. Press, 1977); A. Jdanko, "A Cybernetic Systems Approach to Qualitative Foundations of Social Evolution Considered as a Part of the Universal Evolution," *Proc. 12th Int'l, Congress on Cybernetics* (Namur, Belgium, 1989): 1286–1297.
9. P. Davies, *The Cosmic Blueprint* (Simon & Schuster, 1988).
10. V. I. Krinsky, *Self Organization* (Springer-Verlag, 1984); L. L. Whyte, A. G. Wilson and D. G. Wilson, *Hierarchical Structures* (Elsevier, 1969).
11. V. Turchin, *op. cit.*, ref. 8.
12. R. J. Field, "Chemical Organization in Time and Space," *American Scientist*, 73 (1985): 142–150.
13. W. C. Dampier, *A History of Science* (Cambridge Univ. Press, 1949).
14. S. A. Kauffman, "Self-Organization, Selective Adaptation, and its Limits," in *Evolution at a Crossroads*, editors D. J. DePew and B. H. Weber (MIT Press, 1985): 169–207.
15. H. Haken, editor, *Dynamics of Synergetic Systems* (Springer-Verlag, 1986); J. A. S. Kelso and H. Haken, "New Laws to be Expected in the Organism: Synergetics of Brain and Behavior," in *What is Life? The Next Fifty Years*, edited by M. P. Murphy and L. A. J. O'Neill (Cambridge Univ. Press, 1995): 137–160.

## Chapter 6


1. D. J. De Solla Price, *Little Science, Big Science* (Columbia Univ. Press, 1963).
2. D. Clery, "Element 110 is Created" *Science*, 266 (1994): 1479; R. Koenig, "GSI Bags another New Element" *Science*, 1231 (1996).
3. T. Modis, *Predictions* (Simon & Schuster, 1992).
4. Pierre-P. Grassé, *Evolution of Living Organisms* (Academic Press, 1977).
5. S. Runciman, "The Crusades: A Moral Failure," in *The Crusades*, edited by J. A. Brundage (Heath & Co., 1964): 75–78.
6. T. A. Archer and C. L. Kingsford, "The Crusades: Glory of the Middle Ages," *Ibid.*: 66–68.
7. E. Bradford, *The Sundered Cross* (Prentice Hall, 1967).
8. W. C. Dampier, *A History of Science* (Cambridge Univ. Press, 1949).
9. M. Eigen, *Steps Towards Life* (Oxford Univ. Press, 1992).
10. J. W. Schopf, *Major Events in The History of Life* (Jones and Bartlett, 1992).
11. J. W. Schopf, "The Development of Diversification of Precambrian Life," *Origins of Life*, 5 (1974): 119–135.


## Chapter 7


1. M. W. Zemansky, *Thermodynamics and Statistical Mechanics* (Wiley, 1947).
2. I. Prigogine, *Self Organization in Non-Equilibrium Systems* (Wiley, 1977); G. Nicolis and I. Prigogine, *Exploring Complexity* (Freeman, 1990).
3. F. C. Andrews, *Equilibrium Statistical Mechanics* (Wiley, 1975).
4. E. D. Schneider and J. J. Kay, "Order From Disorder: The Thermodynamics of Complexity In Biology," in *What is Life? The Next Fifty Years*, edited by M. P. Murphy and L. O'Neill (Cambridge Univ. Press, 1995): 161–173.
5. R. Swenson, "Emergent Attractors and the Law of Maximum Entropy Production," *Systems Research*, 6(3) (1989): 187–197.
6. J. S. Wicken, "Evolution and Thermodynamics: A New Paradigm," *Systems Research*, 6(3) (1995): 181–186.


## Chapter 8


1. J. Singh, *Information Theory, Language and Cybernetics* (Dover, 1966); G. Raisbeck, *Information Theory* (MIT Press, 1964).
2. H. J. Morkowitz, *Entropy for Biologists* (Academic Press, 1970): Ch. 10.
3. L. Brillouin, "Physical Entropy and Information," *J. Appl. Phys.*, 22(3) (1951): 338–43; — *Science and Information Theory* (Academic Press, 1962); E. T. Jaynes, "Information Theory and Statistical Mechanics I and II," *Phys. Rev.*, 106 (1957): 620–630 and 108 (1957): 171–190; M. Tribus, *Thermostatics and Thermodynamics* (VanNostrand, 1961); — and E. C. McIrvine, "Energy and Information," *Scientific American*, 225(3) (1971): 179–188.

4. D. Brooks and E. O. Wiley, *Evolution as Entropy* (Univ. Chicago Press, 1986 and 1988).
5. R. U. Ayres, *Information, Entropy, Progress* (AIP Press, 1994).
6. S. Watanabe, *Pattern Recognition: Human and Mechanical* (Wiley, 1985)
7. L. Gatlin, *Information Theory and The Living System* (Columbia Univ. Press, 1972).
8. J. Marx, "Chromosome Yield New Clue to Pairing in Meiosis," *Science*, 273 (1996): 35–36.
9. R. Nowak, "Mining Treasures from Junk DNA," *Science*, 263 (1994): 608–610.
10. H. Quastler, *The Emergence of Biological Organization* (Yale Univ. Press, 1964).
11. M. Eigen, *Steps Towards Life* (Oxford Univ. Press, 1992).
12. J. M. Smith, "Evolutionary Progress and Levels of Selection," in *Evolutionary Progress* edited by M. H. Nitecke (Chicago Univ. Press, 1988: 219–230.
13. H. Moravec, "Human Culture: A Genetic Takeover Underway," in *Artificial Life*, edited by C. G. Langton (Addison Wesley, 1989): 167–199.
14. A. I. Waldrop, "Artificial Intelligence: Introduction to the World," *Science*, 223 (1984): 802–805; G. Kolata, "Order Out of Chaos in Computer," *Ibid.*, 917–919.
15. Q. G. Schwartz and G. J. Klir, "Fuzzy Logic Flowers in Japan," *IEEE Spectrum*, (July, 1992): 32–35.
16. D. Z. Anderson, *Neural Information Processing Systems (*Am. Inst. Physics, 1988); L. Roberts, "Are Neural Nets Like The Human Brain?," *Science*, 243 (1989): 481–482.
17. P. Messina and A. Murli, *Practical Parallel Computing* (Wiley, 1991); D. H. F. Freedman, "A Painless Route to Parallel Computing," *Science*, 253 (1991): 1209–1210.
18. P. J. Denning and W. F. Tichy, "Highly Parallel Computation," *Science*, 250 (1990): 1217–1222; G. Bell, "Ultracomputers: A Teraflop Before its Time," *Science*, 256 (1992): 64.
19. P. Smolensky, "Information Processing in Dynamical Systems," in *Parallel Distributed Processing*, Vol. 1, edited by D. F. Rumelhart and J. L. McClelland (Bradford-MIT Press, 1986): 194–281.
20. G. N. Sardis, "An Integrated Theory of Intelligent Machines by Expressing the Control Performance as Entropy," *Control Theory and Advanced Technology*, 1 (1985): 125–138; Y. S. Abu-Mustafa, "Connectivity versus Entropy," in *Neural Information Processing Systems*, edited by D. Z. Anderson (AIP Press, 1988): 1–8.
21. H. R. Branson, "A Definition of Information From the Thermodynamics of Irreversible Processes," in *Information Theory and Biology*, edited by H. Quastler (Univ. of Illinois Press, 1953): 25–40.
22. J. A. Wilson, "Entropy of Biological Systems, Entropy Not Negentropy," *Nature*, 219 (1968): 534–536.

23. J. S. Wicken, "Evolution and Thermodynamics: The New Paradigm," *Systems Research*, 6(3) (1989): 181–186.
24. K. Haefner, *Evolution and Information Processing Systems* (Springer-Verlag, 1992); P. T. Saunders and M. W. Ho, "On the Increase of Complexity in Evolution," *J. Theor. Biol.*, 90 (1981): 515–530.
25. B-O. Küppers, *Information Theory and the Origin of Life* (MIT Press, 1990).
26. J. E. Gray and A. Vogt, "Information and Meaning," WESScomm, 4, (December, 1994): 2–10; R. Fischer, "From 'Transmission of Signals' to 'Self-Creation of Meaning' Transformations in the Concept of Information," *Cybernetica*, 36 (1993); 229–243.
27. E. Mayr, "How Biology Differs From the Physical Sciences," in *Evolution at a Crossroads*, edited by D. J. DePew and B. H. Weber (MIT Press, 1985): 43–63.
28. E. Culotta and D. E. Koshland Jr., "DNA Repair Enzymes," *Science*, 266 (1994): 1926–1929.
29. J. H. Campbell, "An Organizations Interpretation of Evolution," in *Evolution at a Crossroads*, edited by D. J. DePew and B. H. Weber (MIT Press, 1985): 133–167.
30. D. R. Brooks, D. D. Cummings and P. H. LeBlond," Dollo's Law and the Second Law of Thermodynamics," in *Entropy, Information and Evolution*, edited by B. H. Weber, D. J. DePew and J. D. Smith (Cambridge-MIT Press, 1988): 189–224.
31. J. W. Schopf, *Major Events in The History of Life* (Jones & Bartlett, 1992).
32. J. M. Smith and E. Szathmary, *The Major Transitions in Evolution* (Freeman, 1995).
33. K. Haefner, *Evolution of Information Processing Systems* (Springer-Verlag, 1992): Ch. 1.
34. A. V. Jdanko, "On an Evolutionary-Entropic Interpretation of Information," *Kybernetes*, 23 (1994): 44–57.
35. A. Jdanko, "A Cybernetic Systems Approach to Qualitative Foundation of Social Evolution Considered as a Part of The Universal Evolution," *Proc. 12th International Congress on Cybernetics* (Namur, Belgium, 1989): 1286–1297.
36. A. V. Jdanko, "An Evolutionary Cybernetics Theory of Knowledge and Cognition," *Cybernetics and Systems*, 22(6) (1991): 617–629.
37. D. P. Barlow, "Gametic Imprinting in Mammals," *Science*, 270 (1995): 1610–1613.
38. M. Gardner, "Notes of a Fringe-Watcher," *Skeptical Inquirer*, 20 (1996): 7–9.


## Chapter 9


1. C. R. Woese, "The Emergence of Genetic Organization," in *Exobiology*, edited by C. Ponnamerpuma (N. Holland, 1972): 301–341.
2. K. Haefner, "Evolution of Information Processing-Basic Concept," in *Evolution of Information Processing Systems,* edited by K. Haefner (Springer-Verlag, 1992): 1–46.

3. D. Brooks and E. O. Wiley, *Evolution as Entropy*, (Univ. Chicago Press, 1986 & 1988).
4. D. Layzer, *Cosmogenesis* (Oxford Univ. Press, 1990); — "Growth and Order in the Universe," in *Entropy, Information, and Evolution*, edited by B. H. Weber and D. J. DePew (MIT Press, 1990): 23–39; S. Frautschi, "Entropy in an Expanding Universe," *Ibid.*: 11–22.
5. J. T. Fraser, N. Lawrence and D. Park, *The Study of Time IV* (Springer-Verlag, 1981).
6. W. C. Dampier, *A History of Science* (Cambridge Univ. Press, 1949).
7. A. Linde, "Particle Physics and Inflationary Cosmology," *Physics Today* (Sept. 1987): 61–68; M. M. Waldrop, "Bubbles Upon the River of time," *Science*, 215 (1982): 1082–1083; M. Bartusiak, "Before the Big Bang: The Big Foam," *Discover Magazine* (Sept., 1987): 76–83.
8. S. Hawkings, *A Brief History of Time* (Bantam, 1988).
9. H. Quastler, *The Emergence of Biological Organization* (Yale Univ. Press, 1964).
10. L. L. Gatlin, *Information Theory and the Living System* (Columbia Univ. Press, 1972).
11. W. J. Zurek, *Complexity, Entropy and the Physics of Information* (Addison-Wesley, 1990).
12. P. T. Saunders and M. W. Ho, "On the Increase of Complexity in Evolution," *J. Theor. Biol.*, 63 (1976): 375–384.
13. H. J. Morowitz, *Entropy for Biologists* (Academic Press, 1970); F. C. Andrews, *Equilibrium Statistical Mechanics* (Wiley, 1975).
14. I. Prigogine, *Self Organization in Non-Equilibrium Systems* (Wiley, 1977).
15. H. A. Simon, "The Architecture of Complexity," *Proc. Amer. Phil. Soc.*, 106 (1962): 467–482.

## Chapter 10

1. I. H. Campbell et al., "Synchronism of the Siberian Traps and the Permian-Triassic Boundary," *Science*, 258 (1992): 1760–1763.
2. D. H. Erwin, *The Great Paleozoic Crisis* (Columbia Univ. Press, 1993).
3. R. A. Kerr, "The Greatest Extinction Gets Greater," *Science*, 262 (1993): 1370–1371; K. A. McDonald, "Claim that Asteroid Moved Continents," *Chronicle of Higher Education* (1/20/93): A6–A8.
4. C. C. Swisher et al., "Coeval $^{40}$Ar/$^{39}$AR Ages of 65.0 Million Years Ago From Chicxulub Crater," *Science*, 257 (1992): 954–958; D. M. Raup and D. Jablonski, "Geography of End-Cretaceous Marine Bivalve Extinctions," *Science*, 260 (1993): 971–973; R. A. Kerr, "High Impact Tied to Mass Extinction," *Science*, 257 (1992): 878–880.
5. J. D. Archibald, "The Importance of Phylogenetic Analysis ... Mammals in North America," *Paleobiology*, 19(1) (1993); 1–27.
6. M. R. Rampino and R. B. Stothers, "Flood Basalt Volcanism During the Past 250 Million Years," *Science*, 241 (1988): 663–668; M. R. Rampino

and B. M. Haggerty, "The 'Shiva Hypothesis': Impacts, Mass Extinction, and the Galaxy," *Earth, Moon, and Planets*, 72 (1996): 441–460; B. A. Kerr, "Origins of Extinctions," *Science*, 257 (1992): 486–487.
7. C. Ponnamperuma, *Exobiology* (N. Holland, 1972).
8. D. A. Russel and R. Sequin' "Reconstruction of a Small Cretaceous Theropod, ..., A Hypothetical Dinosauroid," *Syllogeus*, 37 (1982): 1–43.
9. F. Jacob, "Evolution and Tinkering," *Science*, 196 (1977): 1161–1166.
10. W. Day, *Genesis on Planet Earth* (Yale Univ. Press, 1984).
11. M. M. Kamshilov, *Evolution of the Biosphere* (MIR, 1974, 1976).
12. J. Weiner, "Evolution Made Visible," *Science*, 267 (1995): 30–33.
13. N. Weiner, *Cybernetics of Control and Communication in the Animal and the Machine* (MIT Press, 1977).
14. J. F. Crow, "Mechanisms and Trends in Human Evolution," in *Evolutionary Anthropology*, edited by H. K. Bleibtreu (Allyn & Bacon, 1969): 273–287.
15. L. Margulis, "Symbiosis and Evolution," *Scientific American*, 225(2) (1971): 48–61; J. E. Lovelock, *A New Look at Life on Earth* (Oxford Univ. Press, 1987).
16. R. W. Burhoe, "War, Peace, and Religion's Biocultural Evolution," *Zygon*, 21 (1986): 439–472.
17. R. A. Kerr, "Exobiology: SETI Faces Uncertainty on Earth and in Stars," *Science*, 258 (1992): 27.
18. M. D. Popaagiannis, *The Search for Extraterrestrial Intelligence* (Reidel, 1985).
19. G. Zorpette, "Radio Astronomy — New Window on Universe," *I.E.E.E. Spectrum* (Feb. 1995): 18–25.
20. J. Glanz, "Computer Scientists Rethink Their Discipline's Foundations," *Science*, 269 (1995): 1363–1364; P. D. DiVincenzo, "Quantum Computation," *Science*, 270 (1995): 255–261.
21. G. Stock, *Metaman* (Simon & Schuster, 1993); J. C. Glenn, *Future Mind: Artificial Intelligence* (Acropolis Books, 1989).
22. Special issue on "Bioelectronic Vision," *I.E.E.E. Spectrum*, May 1996, vol. 33, no. 3.
23. M. W. Zemansky, *Thermodynamics and Statistical Mechanisms* (Wiley, 1947).
24. Moravec, "Human Culture: A Genetic Takeover Underway," in *Artificial Life*, edited by C. G. Langton (Addison Wesley, 1989): 167–199; — *Mind Children* (Harvard Univ. Press, 1989).
25. R. A. Brooks, "New Approaches to Robotics," *Science*, 253 (1991): 1227–1232; F. Flam, "Swarms of Mini-Robots," *Science*, 257 (1992): 1621; M. M. Waldrop, "Fast, Cheap, and Out of Control," *Science*, 248 (1990): 959–961.
26. Special section: "Engineering a Small World: From Atomic Manipulators to Microfabrication," *Science*, 254 (1991): 1300–1341; C. Schneiker, "Nanotechnology With Feynman Machines," in *Artificial Life* edited by C. G. Langton (Addition-Wesley, 1989): 443–500; E. Drexler, "Biological Nanomechanical Systems," *Ibid.*: 501–519.

27. A. G. Cairn-Smiths, *Seven Clues to Origin of Life* (Cambridge Univ. Press, 1985).
28. H. Hurvitz, Private communication.
29. M. Eigen, *Steps Towards Life* (Oxford Univ. Press, 1992).

## Appendix

1. J. M. Smith, *Mathematical Ideas in Biology* (Cambridge Univ. Press, 1968); E. O. Wilson and W. H. Bossert, *A Primer of Population Biology* (Sinauer Assoc., 1983); I. Guttman and S. S. Wilks, I*ntroductory Engineering Statistics* (Wiley, 1965).

# INDEX